Just-In-Time Manufacturing

A Practical Approach

Arnaldo Hernandez

Prentice Hall
Englewood Cliffs, New Jersey

Library of Congress Cataloging in Publication Data

Hernández del Campo, Arnaldo.
Just-in-time manufacturing.

Bibliography: p.
Includes index.
1. Just-in-time systems. I. Title.
TS155.H395 1989 658.5'6 88-32482
ISBN 0-13-514027-7

Editorial/production supervision and
interior design: Tally Morgan
Cover design: Ben Santora
Manufacturing buyer: Mary Ann Gloriande

This book can be made available to business
and organizations at a special discount when
ordered in large quantities. For more information
contact:

Prentice-Hall, Inc.
Special Sales and Markets
College Division
Englewood Cliffs, N.J. 07632

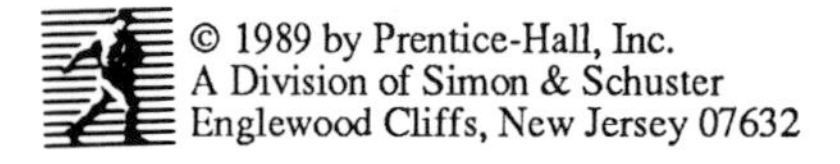

Printed in the United States of America

10 9 8 7 6 5 4 3 2 1

ISBN 0-13-514027-7

Prentice-Hall International (UK) Limited, *London*
Prentice-Hall of Australia Pty. Limited, *Sydney*
Prentice-Hall Canada Inc., *Toronto*
Prentice-Hall Hispanoamericana, S.A., *Mexico*
Prentice-Hall of India Private Limited, *New Delhi*
Prentice-Hall of Japan, Inc., *Tokyo*
Simon & Schuster Asia Pte. Ltd., *Singapore*
Editora Prentice-Hall do Brasil, Ltda., *Rio de Janeiro*

To my wife Barbara,
for a love that is always just-in-time

Contents

Preface

Just-In-Time is a new way of thinking about manufacturing. The challenge of Just-In-Time is that it constitutes a complete departure from the old operational systems that have been used for many years in our factories. As you progress in reading this book, you will see that Just-In-Time is not a narrow discipline applied to specific areas in manufacturing. Instead, Just-In-Time is a collection of disciplines that reaches a wide range of departments that normally are not considered operational in nature. I have purposely covered the impact of Just-In-Time on departments that are usually not included in other books or seminars on the subject. This book will give the reader a practical road map to applying Just-In-Time; it starts with the basic concepts, then covers the critical operational rules that pertain to all departments involved, and finally gives the specific instructions required to start a Just-In-Time system from scratch. We will see that Just-In-Time is a two-dimensional process and that it crosses many departmental boundaries and reaches all of the many levels of workers and management.

Most of the Just-In-Time concepts have been developed in Japan through many years of hard work and attention to detail. These concepts evolved, in many cases within different companies, toward the basic principle that manufacturing products should be done in the most efficient way and in the minimum amount of time, result in the highest possible quality, and use the minimum quantity of raw materials at hand. As this evolution occurred, the concept of waste as any activity that adds no value to a product took a new meaning with respect to the production of industrial goods. This book will present all the pertinent concepts as they apply to the different departments which in some way influence the manufacturing process.

This book discusses how Just-In-Time functions at all levels in an organization. Just-In-Time requires constant attention, not only from workers but also from all levels of management. This book will also present a practical approach to implementing the system from the beginning, and it will guide the reader through the different steps required to implement the system companywide.

Another aspect of Just-In-Time presented in this book is the notion of total quality control (TQC). No Just-In-Time system implementation could succeed without a TQC system. Quality is of the utmost importance in a Just-In-Time system, for one of the first things that Just-In-Time does is to eliminate excess parts not needed to build products. Just-In-Time forces manufacturers to make sure that the parts at hand are good and meet quality standards.

Finally, it should be said that Just-In-Time is a critical discipline that can help address the industrial challenge most American manufacturers are facing today. Everyone is aware of the competitive nature of the today's industrial markets. While in the past American products had no competition in American markets, this is no longer true; our companies are now engaged in a fierce contest with companies from all over the world, particularly from Japan. This new challenge is forcing American manufacturers to reassess their organizations in order to produce products of higher quality while reducing operating costs. Doing this requires revamping manufacturing organizations to make them leaner and more productive.

The challenge appears even greater when we consider that most of our manufacturing organizations are loaded with large inventories, have clumsy operational environments, generate massive amounts of paperwork, and use very poor methodologies to solve problems when they occur. All these factors have turned American manufacturing organizations into large dinosaurs that have terrible response times and high operating costs and produce poor quality products.

This book offers a practical way to solve those problems by going back to basics. The principles and systems explained here are clear and ordered in such a way that any organization with the desire to change can implement them and obtain immediate results.

In this book, the Just-In-Time concepts are developed theoretically rather than by presenting case studies of companies. However, visits to companies implementing Just-In-Time and long experience implementing Just-In-Time myself have helped me to understand these concepts and be able to present the Just-In-Time system clearly.

This book is about common sense and how we can make any manufacturing system operate in the simplest and most productive way and at the same time produce the highest-quality goods.

Arnaldo Hernandez

Acknowledgments

First, I want to express my gratitude to Paul W. Becker, Senior Acquisitions Editor and Assistant Vice President at Prentice-Hall. I've known Paul for over fifteen years, ever since I taught a graduate course in computer sciences at Stevens Institute of Technology. At that time, we often talked of writing a book about computer systems. That book was never realized, but throughout the years I have felt that I owed Paul a book. I hope this one will settle my perceived debt. Paul has been very supportive and has offered many suggestions and comments.

I want to thank the president of Metaphor Computer Systems, Donald J. Massaro, for his support of and commitment to the Just-In-Time program at Metaphor. Also, I want to thank Metaphor's manufacturing managers: Len Andes, Director of Materials; Dan Denver, Director of Purchasing; Terry Farlow, Quality Assurance Manager; Ron Jones, Manufacturing Manager; Marty Martinez, Industrial Engineering Manager; and Bob Franchini, Director of Manufacturing Support. These individuals have supported Metaphor's Just-In-Time effort and contributed many practical ideas for improving the system. They also read this manuscript and contributed many suggestions.

I also want to acknowledge the great support given to Metaphor by Hamilton/Avnet's In-Plant Store Program. Credit belongs to Mike Burke, Hamilton's In-Plant Store Program manager at the time, and to Ron H. Mabry, Hamilton/Avnet General Manager.

Finally, I want to thank my wife, Barbara, for her unconditional support and immense patience during the writing of this book. Her encouragement of my writing projects and her protection of my writing time has been invaluable. I couldn't have done it without her inspiration and love.

The Way of Strategy

Do not think dishonestly.
The Way is in training.
Become acquainted with every art.
Know the Ways of all professions.
Distinguish between gain and loss in worldly matters.
Develop intuitive judgement and understanding for everything.
Perceive those things which cannot be seen.
Pay attention even to trifles.
Do nothing which is of no use.

Miyamoto Musashi (1584–1645)
A Book of Five Rings

1 ★

The Need for Just-In-Time

Just-In-Time manufacturing is an extension of the original concept of managing the material flow in a factory to reduce the inventory levels. In fact, there is much more involved in a manufacturing organization than reducing inventories to control costs. Manufacturing has to deal with other issues, such as process control, level of automation, flexible manufacturing, machine setup times, direct labor productivity, overhead, supplier management, engineering support, and the quality of the product delivered to customers.

A modern manufacturing organization has to deal efficiently with these issues in order to operate a smooth, productive, and quality-minded department. This book presents a fresh approach to deal with them simply and pragmatically. It describes techniques long practiced by well-established Japanese companies and by some progressive American companies. Included in the book is knowledge gained at Metaphor Computer Systems during the process of applying these techniques. I hope to help the newcomer by offering a complete blueprint for the Just-In-Time strategy. I also hope to offer the current practitioners new areas of exploration so they can reach higher levels of accomplishment. This book is essentially about returning to common sense and simplicity in manufacturing. It is also about survival.

1.1 MANUFACTURING STRATEGY FOR SURVIVAL

Manufacturing is no longer a local matter. Advances in communication and transportation have greatly reduced the world's size, and manufacturing should now be considered a world affair. The consequent variety of choices make decisions regarding manufacturing strategy very difficult and risky. Today American companies face world competition, and manufacturing is at the heart of the problem.

To maintain their competitive edge, companies engaged in manufacturing products face the difficulty of reducing costs and improving their quality levels. One way to accomplish these goals is to reduce the labor and material cost required to build the product. These are the obvious factors usually considered, but they often don't reflect the complete picture. Included in the cost equation should be the overhead associated with the process of building a product. This overhead in some instances is what tips the balance to a particular implementation side.

For the past few years, American companies have faced two choices concerning where they should build their products. Then, after struggling to define their strategies, many of these companies followed the cheap-labor-and-materials route. They are high-volume producers and they moved their manufacturing plants to places where labor rates are very low in comparison with U.S. rates.

Many others were inclined to institute automation in order to keep their factories competitive in the United States. These companies, however, faced the heavy financial commitment required to procure the automation equipment that would reduce the labor content of their products. Unfortunately, sometimes companies were unable or unwilling to commit the requisite large capital investment. These companies commonly missed the point that automation is only part of the solution to the cost reduction of a product, for labor is responsible for only a small portion of the product's cost.

What is of most importance is to use the correct strategy in manufacturing. Most companies have a product strategy and a marketing and sales strategy, but they do very poorly in developing a manufacturing strategy. When these companies develop a product and introduce it in the market against the competition, they fail because the cost is too high, they cannot produce the volume required, or their quality levels are unacceptable.

We want to stress in this chapter the need to develop a manufacturing strategy to complement product and market strategies. Without all three strategies, any company would be handicapped in its quest for market dominance and would probably be doomed to failure. It is necessary to develop a commitment to manufacturing earlier in the product's development phase. It is important to use common sense in studying the different choices and to carry out decisions that will make the manufacturing process effective, fast, and burdened with very low overhead. That is what Just-In-Time manufacturing is all about.

1.2 THE COST OF BUILDING A PRODUCT

Products built in a manufacturing organization have three cost variables: materials, labor, and overhead. The materials variable is the cost of the material used to build the product. The labor variable is the hours invested in assembling and testing the product. Overhead is everything required to support the factory while it builds the product. Overhead includes the rent of the building, the interest payments to the bank for the equipment bought to build the product, and the cost of money invested in the inventory.

With a few exceptions, the material content of the product is the biggest portion of the cost of a product. The second is overhead, with direct labor the smaller of the three. In manufacturing, all three variables have to be managed in order to achieve the lowest cost without compromising the quality of products delivered to the customers.

Just-In-Time addresses all three cost variables in a similar way: understand them, reduce costs by using common sense and simple procedures, and cut out all that is not necessary.

Material and Labor Content

All products have some ratio content of material, labor, and overhead. The material content of a product is the sum of the cost of all the parts used to build the product. Most companies list on their computers all the parts required to build their products. Each part has a standard cost associated with it. This is the parts cost agreed to by the materials and cost accounting departments. The roll-up of the cost of these parts determines the standard cost of the product.

For the labor content of the product, one could go around the factory and locate the people who build the product. The labor cost is calculated by adding up all the time that those people invest in assembling and testing the product and then multiplying by the average hourly salary.

The ratio of material and labor costs in a product varies depending on the type of product. With very few exceptions, materials make up the larger share, usually 70 to 90 percent of the total cost.

In a typical manufacturing organization, the arrival and flow of parts on the factory floor occurs with a certain degree of randomness. There are slippages on the suppliers' production and shipping schedules. There are delays as a result of not ordering the parts with the proper lead times. There are quality problems which prevent using all the parts that have been received. Finally, there are yield variations, with fewer products made than initially planned. In order to synchronize all these variables, manufacturers use buffer inventories as a safety measure to ensure that parts are available when needed. Then, as a result of this safety-prone mentality, buffer inventories sprout and grow practically everywhere in the manufacturing process to cover for the above deficiencies. This approach is quite inefficient. About 70–90 percent of the product cost in material is stored for no other purpose than to be there Just-In-Case.

Just-In-Time calls for the reduction or elimination of all buffer inventories to save the cost associated with carrying the material and to avoid the possibility of material obsolescence. In a Just-In-Time system, there needs to be aggressive attempts to solve manufacturing problems so as to obtain the maximum yield of the processes and the materials used. Just-In-Time forces people to act on problems right away, for no additional material is available as a safety. Having no buffer inventories also forces manufacturers to work closely with suppliers in order to assure timely and high-quality deliveries. Manufacturers no longer have the luxury of covering for late deliveries or poor quality. Just-In-Time puts manufacturers in a position where they don't have any other alternative than to solve problems as soon as they occur. They have their backs against the wall.

Investment of Money in Manufacturing

In a business cycle, companies invest large amounts of money in the materials required to build products. Then they recover the money by selling the products to the customers. There are other areas where companies also spend large sums of money, but it takes a longer time to recover the investment. One example is the investment in the capital equipment required to get a production line off the ground. A company recovers this investment through the process of using the equipment to build the product. Accountants call this process a return on investment.

Normally, the finance department would ask the manufacturing manager for a return on investment analysis before the company commits itself to a major capital equipment investment. The manufacturing manager can sometimes produce this type of analysis very easily, but other times it might be hard to justify the investment. A typical example is an investment in a material-handling system for the production floor.

Any production line needs to move material around the factory. The movement of material to work centers directly affects the productivity gains. However, an increase in efficiency is sometimes not enough to produce a clear return on investment—unless the high volume of the line makes it obvious. It is important for management to weigh the intangibles of the investment in capital, for example, the savings achieved in moving the material efficiently with the corresponding gains in workers' productivity and in high-quality outputs. These factors can be called "gut feeling factors." It is the job of a manufacturing manager to point them out to the finance department during the justification process.

A good Just-In-Time strategy must include a blueprint for capital equipment investment. It should include not only the capital equipment required to build and test the products but also the equipment required to move material efficiently. Just-In-Time systems are very dynamic in moving material, and considerable time should be spent planning the most efficient way to accomplish this task.

Besides capital investment, there is another area requiring careful planning before any commitment to capital: the level of inventories required to build products. Of course, there must be some "level of inventories," because it is impossible to build a product with no inventory at hand. Even in cases where a manufacturer perfectly tuned every supplier's delivery, where there are no quality or process problems, and where the factory is completely linear, the manufacturer would need at least the material required to fill the production line during the time the product is being made. This is what is called the work-in-process (WIP) inventory.

1.3 MANUFACTURING STRATEGY

In general, developing a strategy is a two-step process. First, a set of goals is defined. This step is equivalent to looking at a map and marking a destination. Second, the means of achieving the goals are defined. This is equivalent to selecting the roads for getting to the destination.

A manufacturing strategy must answer all the key questions related to the implementation of the factory and the operating environment. The strategy must define the factory capacity, level of inventories, quality goals, level of automation, and operational environment.

These strategic decisions will affect how manufacturing deals with four critical operational areas: suppliers, materials management, factory process, and customers. If these areas are not well planned and coordinated, they could render the manufacturing strategy useless.

Throughout this book are described the different pieces of a manufacturing strategy that will provide the framework for a Just-In-Time strategy.

1.4 JUST-IN-TIME STRATEGY

Most people associate a Just-In-Time system with an inventory reduction program or with the goal of having zero inventory. The truth is that Just-In-Time is much broader and affects the operation of many departments in a company. On the other hand, inventory reduction *is* one of the key goals of a Just-In-Time system, and it certainly is one of the results of using Just-In-Time techniques. Just-In-Time is a system designed to make manufacturing organizations operate efficiently and with a minimum of human and mechanical resources. Just-In-Time also improves quality, reduces inventory levels, and provides maximum motivation to solve problems as soon as they occur.

Using again the analogy between developing a strategy and picking a destination and a route on a map, Just-In-Time is analogous to the set of roads that will take someone to his or her destination in the most efficient and straightest way.

Just-In-Time is simplicity, efficiency, and minimum waste.

Just-In-Time and Waste

Just-In-Time introduces a new definition of waste in manufacturing. Waste is usually taken to be material scrap, rework, or line fallout. Just-In-Time describes as waste anything that is not necessary for the manufacturing of the product or is in excess, for example, buffer inventories to cover for defective parts in the production lines or nonlinear building rates, labor hours spent producing products that are not necessary, labor hours spent reworking products because of poor quality or engineering rework orders, and time invested in setting up machine tools before they start processing parts. All this wasted time and material increases the cost of the product and lowers its quality. Just-In-Time is a crusade to eliminate all forms of waste. It is also a drive to simplify the manufacturing process in order to quickly detect problems and force immediate solutions.

Definition of Just-In-Time

You should have a general idea of what Just-In-Time is all about. Consider now one of the many definitions that are found in the papers and books on the subject.

Just-In-Time can be defined as a production system designed to eliminate waste in the manufacturing environment. As mentioned above, waste is anything that doesn't contribute directly to the value of the product.

One way to expand on this definition is as follows: In a Just-In-Time system, the necessary materials are brought to the necessary place to build the necessary products at the exact time when they are required.

After you have read this definition, you are probably going to say, "Wait a minute! This is all a dream. Murphy's Law is also applicable to a manufacturing organization." You are correct, but one of the benefits of a Just-In-Time system is to reduce the effects of Murphy's Law by removing as much as possible the factors that allow it to operate.

In addition to the Just-In-Time definition, there are two rules which are required for the implementation of a Just-In-Time system.

First, use only high-quality parts and processes. Just-In-Time calls for minimum buffers in materials and subassemblies. Thus, when the moment to build the product arrives, the parts used in the production process must be the best available. This rule will ensure high yields and predictability on the production line.

The second rule has to do with the lot size of the product being built. The Just-In-Time ideal lot size is one. Thus, the second rule is this: Always strive to build the smallest lot size of any product, regardless of the production volume of the product.

These two rules are the pillars of Just-In-Time's operating principles. A violation of either of them would cause serious problems in the implementation of the system.

1.5 JUST-IN-TIME AND THE COMPETITION

The Just-In-Time system is the second shock wave to jolt U.S. industry. The first one, of course, occurred when the Japanese competition proved that their products were of better quality than ours, particularly in such industries as automobiles and electronics.

Right after the first shock wave, U.S. industry sent hordes of technicians to Japan to figure out how the Japanese were producing products of such high quality. At the time, they discovered the value of the *quality circles*. Soon many American companies embraced the concept, only to abandon it later when they didn't obtain the same results as their Japanese counterparts.

The lesson is that quality circles were more than a procedure to improve quality. They involved a total commitment to manufacture products according to very high standards. Americans saw the process only as a procedure and they failed as a result. It wasn't until later that they realized that not only were the procedures of great importance, but so was a true commitment on the part of the management and the workers to make the system work. Quality required involvement at all levels of the company.

With Just-In-Time the same problem recurs. Most U.S. companies are trying to embrace the concept now, thinking that it will be the cure for all their ills. But before they go any further, they must realize that Just-In-Time involves an all-out war against waste in any shape or form. Also, a Just-In-Time system will only succeed when management, workers, and suppliers make a strong commitment to work together in solving the problems associated with it Partial commitment or top management noninvolvement will result in failure.

At this point, your first thought is probably that this commitment will be very difficult to achieve, for traditionally management, workers, and suppliers have been considered to be adversaries and to have conflicting goals. One of the first objectives of a Just-In-Time system is to change this perception and turn their relationships into partnerships, set common goals, and create win-win situations. This is particularly true regarding the relationship between manufacturers and suppliers, for manufacturers will have to make long-term commitments to suppliers and will depend very heavily on their performance in order to meet production goals.

The beginning of Just-In-Time can be traced back to Toyota in the years after World War II. At that time, Toyota was near bankruptcy; as the result, management started a crusade to improve the company's productivity and the quality of its goods. It took Toyota many years to implement this goal, and the process affected all the levels in the company. But the effort created one of the most efficient companies ever.

The Toyota way of operation evolved into the Toyota system. Many papers described the system and many people studied Toyota's astounding success in reducing inventory levels and improving quality.

One of the concepts developed under the Toyota system was to move materials to work centers (1) in a continuous flow rather than in a batch mode, (2) in the smallest possible quantities, and (3) only when the parts were necessary to build products. Toyota also became very efficient in reducing the setup time of dies in their press machines. This gave their work centers the flexibility to process small production lots very efficiently.

The Toyota system also applied other commonsensical concepts, such as the Kanban system, where the materials are moved on the floor in a pull fashion and controlled by a card called a Kanban. Toyota also gave authority to the workers to stop the production line whenever something went wrong in the production process. Finally, they automated many tasks with machines requiring very little supervision from operators. These machines would stop when something went wrong, allowing one worker to operate many machines without compromising the quality of the parts.

In summary, the Toyota system evolved into the Just-In-Time system. This is especially clear with regard to the emphasis on eliminating waste in all the aspects of the manufacturing process and with regard to the definition of waste as anything that does not add any value to the product.

1.6 PRODUCT QUALITY AND CUSTOMER SATISFACTION

A factory can be conceived of as a large machine that has many knobs for adjustment. Just-In-Time is the adjustment procedure that tunes the machinery for optimum performance. The result of this tuning is that the machine produces the best products that can be built and yet operates with a minimum amount of fuel.

An important benefit of tuning a factory process for optimum performance is that the work centers will produce products of the highest possible quality. Note the use of the word "possible." No factory can build a high-quality product if the raw materials used to build it are of low quality or if the design of the product does not meet the desired specifications.

Quality is everybody's business, and we must work very hard to ensure that everybody shares the responsibility of producing a first-class product.

In a Just-In-Time system, quality is a critical requirement. The system allows no fat to compensate for quality problems on the line. Conversely, Just-In-Time forces the manufacturing organization to fix quality problems as they occur; otherwise, the line will not continue producing. Poor quality parts become very obvious in a manufacturing process using Just-In-Time principles, for there are no buffer parts to replace the bad ones. The production line has only the exact number of parts required to meet the build schedule.

The most important result of a quality system is the increase in customer satisfaction, for the products will meet customers' expectations. Just-In-Time forces manufacturers to implement a quality system. We call such a system a total quality control (TQC) system. Chapters 8 and 10 will present the fundamentals of a TQC system. Figure 1.1 presents a

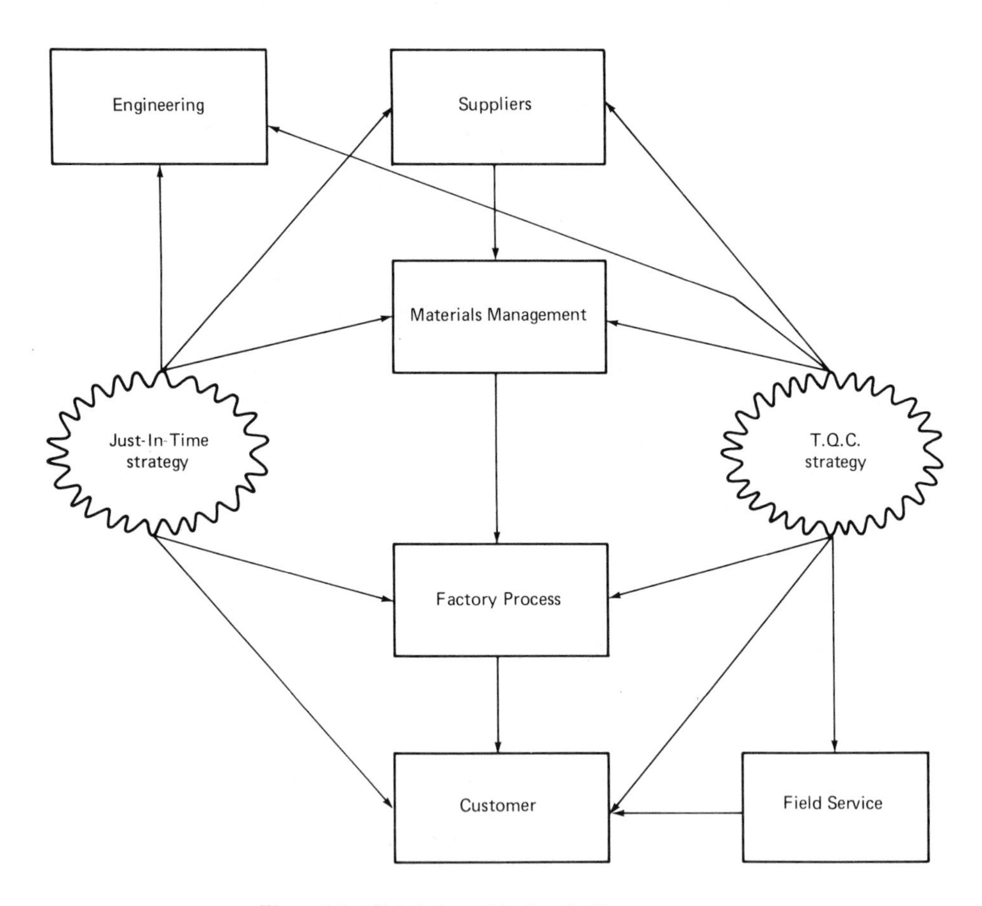

Figure 1.1 Global view of the Just-In-Time system.

global view of the Just-In-Time and TQC systems as they relate to the design, manufacturing, and customer process in a company.

1.7 MANAGEMENT COMMITMENT TO JUST-IN-TIME

Just-In-Time can be thought of as functioning somewhat like a religion: It has its leaders, who inspire and guide, and also its believers, who implement and follow. As with a religion, both groups are necessary.

After instituting a Just-In-Time system, a manufacturing company will have to walk on the edge of disaster for quite some time. Its employees are going to do more with less. They are going to point out defects and problems as soon as they occur. They are going to stop a production line if there is a problem and not start it again until the problem is solved. The company is going to make commitments to single source suppliers for a longer time and refrain from always buying the cheapest parts. All these actions need the support of a solid commitment by management.

Management is a very important force in any company. Their endorsement of the Just-In-Time system is required at the very start. With time, Just-In-Time results will reassure them that the company is doing the right thing. However, before sailing in a smooth ocean, the company is going to face some rough waters. Everyone, including the managers, will have to persevere, even if there are no immediate improvements. In the end, the managers will win by showing that the Just-In-Time system delivers, and the company and its customers will win too.

Commitment at All Levels

One of the problems encountered when talking about implementing a Just-In-Time system is that the manufacturers get scared at the thought of more work and less results. Many manufacturing organizations operate according to old-fashioned ideas that conflict head-on with Just-In-Time principles. For example, in manufacturing it was always assumed that second source suppliers were good for backup. It was also believed that buffer inventories ensured success in meeting production schedules and that the production line must never be stopped except in a matter of life and death. Finally, fixing quality problems was normally postponed for the sake of production output, possibly with the assumption that time and volume would take care of the problem.

Middle management, front-line supervisors, and workers need to embrace the Just-In-Time system totally or it won't work. The question becomes, How does a company make believers of these people? Probably the only way to achieve this goal is to use training and participation. The company should immediately start a training program for the people involved in the program. It is also critical that they participate in the definition and execution of the Just-In-Time goals. This participation is vital to the success of the system. Therefore, the company should not try to save time and money in training. The reward will show later when it gets universal participation in the Just-In-Time system. Chapters 11 and 12 address this issue in detail.

1.8 SUMMARY

There are two things that need to be emphasized before closing this chapter. One is that Just-In-Time is going to change the way manufacturing organizations do things. Some of the Just-In-Time concepts are completely opposite to traditional ways of thinking. It is human nature to resist change, and the implementation of a Just-In-Time system is typical of this. People resist these new ideas and call them risky. But it is important to realize that Just-In-Time will not work if it has to be forced against everybody's will. Voluntary participation is necessary. One sure way to achieve this is to start training very early in the process and to encourage individual participation. Without training and participation, the implementation will be seriously jeopardized.

The second point is that Just-In-Time is not a magic formula that will cure all manufacturing ills. Just-In-Time requires hard work, attention to details, and the use of creative thinking to solve problems in a simple way. Applying Just-In-Time principles is not a rigid, cast-in-concrete process. Just-In-Time is flexible and can be tailored to any manufacturing organization regardless of its product and volume. It is important to note that the Just-In-Time principles do not depend on a particular group of people or a particular environment.

This book offers the reader information and advice based on my experience applying the Just-In-Time concepts within Metaphor Computer Systems. I believe that when the reader finishes this book, he or she will have a practical framework from which to start implementing the concepts. Then it will be up to the reader to implement the necessary modifications to tailor Just-In-Time to the individual situation.

REFERENCES

HAYES, ROBERT H., and STEVEN C. WHEELWRIGHT. *Restoring Our Competitive Edge: Competing through Manufacturing*. New York: John Wiley & Sons, 1984.

PORTER, MICHAEL E. *Competitive Strategy: Techniques for Analyzing Industries and Competitors*. New York: The Free Press, 1980.

SCHONBERGER, RICHARD J. *Japanese Manufacturing Techniques: Nine Hidden Lessons in Simplicity*. New York: The Free Press, 1982.

———. *World Class Manufacturing: The Lessons of Simplicity Applied*. New York: The Free Press, 1986.

SKINNER, WICKHAM. *Manufacturing: The Formidable Competitive Weapon*. New York: John Wiley & Sons, 1985.

SUZAKI, KIYOSHI. "Japanese Manufacturing Techniques: Their Importance to U.S. Manufacturers." *Journal of Business Strategy*, 5, No. 3 (Winter 1985).

2 ★

What Just-In-Time Means to Materials Management

Inventory is one of the most important assets that a company owns. Normally, as a company's sales increase, the demand for cash to finance inventory follows the same growth pattern. A Just-In-Time system dedicates a major portion of its attention to manage the inventories throughout the manufacturing organization. It should be pointed out that Just-In-Time doesn't mean zero inventory. Just-In-Time is a set of procedures that are used by the materials department in working with suppliers and with the quality, engineering, and manufacturing departments to reduce as much as possible the use of buffer inventories. Just-In-Time calls for synchronizing the movement of materials throughout the production process in such a fashion that there are very short waits between the different subprocesses. Just-In-Time also moves the materials in the factory based on consumption rather than top-down planning.

2.1 PUSH SYSTEMS

Among the most fundamental changes that Just-In-Time introduces in a manufacturing organization is the institution of a pull system instead of a push system. Push and pull refer to ways of moving materials throughout a factory.

Most companies operate in a push environment. They use master schedules and material requirement planning (MRP) outputs to drive their production schedules and the movement of materials in the factory. A pull system, on the contrary, uses bottom-up demand, which is driven by the consumption rates of parts in the production process.

Materials Planning in a Push System

Factories in general can use two different approaches to plan and build products to fill customer orders. The first approach is to build to order. This means that when an order arrives at the sales department, it creates a factory's demand to manufacture the product according to how the customer wants it. This mode of operation is not usual, and it is normally used when the product ordered has to be customized. The factory operating in this mode will have a long lead time for delivering the product, because the manufacturing only starts after the customer has placed the order. An example would be a semiconductor company building a customized semiconductor designed to meet a customer's specifications.

The other approach is to build to a demand forecast. A demand forecast comes from the sales department, which in turn received it from the sales staff in the field. The sales support staff revise the forecast in light of the company's past sales history. They also review it against the sales plan for the next three to six months to ensure that it matches the company's revenue plan. Once they have completed the adjustment process, the sales department passes the forecast to the materials organization in the manufacturing department. The forecast describes item by item the quantity of products that the field sales staff plan to sell.

The sales operation staff normally consult the finished goods inventory at the time they compare the product demand forecast and the sales forecast. This allows them to plan the amount of finished products required to support the sales effort in the field.

The materials group converts the sales forecast into a manufacturing master schedule and inputs this data into an MRP system. An MRP is a top-down computer program that explodes the product quantities in the master schedule all the way down to the lowest-level parts. It takes into consideration all the inventory locations in the factory and nets the parts required to build the products in the quantities listed on the master schedule.

Figure 2.1 shows the flow of an MRP system. Notice that a downward arrow connects the different operational areas. This means that the build schedule and the materials releases follow the same direction without the proper feedback to allow self-correction in case of changes in the schedules or trouble with the production process or the materials supply to the production line.

The MRP system will report the parts that are on order to cover for the materials that are not at hand but are needed to meet the build schedule. It will also report the parts that must be ordered to fill possible shortages. In addition, it can report the excess parts that are not needed for the build plan and the parts that are obsolete or cannot be used by a higher assembly.

In a push system, the MRP system triggers a series of work orders required to build the products in the quantities specified on the master schedule. A work order is a release used by the materials planner to provide manufacturing with the materials needed to build a certain number of products. The planner will open a work order on the computer authorizing the stockroom to release the parts to manufacturing. We call this process *kit pulling*. The stockroom clerk uses a kit pick list from the computer to pull all the parts required to build

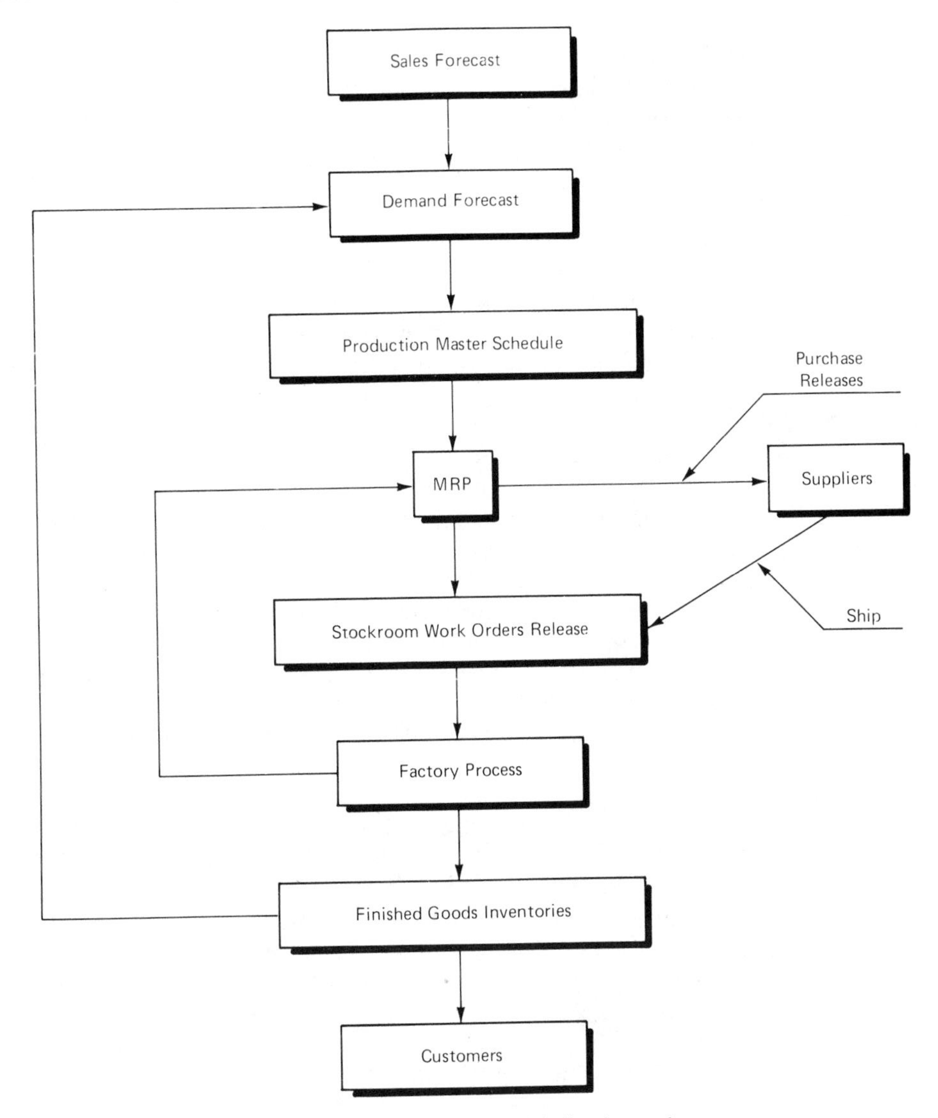

Figure 2.1 Build schedule and materials flow in a push system.

the number of units specified by the work order. When the parts are together, the stockroom clerk releases the kit and the work order paperwork to the manufacturing floor.

A work order is more than a manufacturing release to build the product. A work order is a financial entity in itself. When the planner sends a work order to manufacturing, the workers on the production line will invest a certain number of labor hours in building the product. The work order will track this labor, rolling up its value and comparing it against the standard labor costs previously stored in the computer's data base. This process will tell the manufacturing managers about the efficiency of their organization and the actual cost of building the product.

Work orders are laborious and time-consuming. The materials planner has to open them in the computer and then issue the paperwork to the manufacturing floor, keeping track of the paperwork throughout the life of the production process. After completion, the planner closes the work order on the computer and transfers what has been produced to the stockroom if the product is a subassembly or to the finished goods inventory if it is a final product.

There are many operational problems with work orders, and tracking is one of them. Normally, a manufacturing organization using a work order system has many work orders opened at the same time. It requires a lot of overhead on the part of a planner to open, track, and close the work orders. Also, work orders have the tendency to split. For many operational reasons not important to go into at this moment, planners split work orders into smaller ones during the production process. This split causes more paperwork and more tracking. Also, workers sometimes move materials on the floor in such a way that they mix the parts for one work order with the parts for another. All this tracking and handling takes precious time, creates confusion, and is wasteful.

The MRP system normally has a feedback loop that receives information about the factory work-in-process (WIP) status at the time it is planning the material requirements. The problem is that the response time of the system is too slow. When the schedule corrections reach the manufacturing floor, the excess inventory and product completions are already at hand.

Materials planners also have the tendency to issue work orders to the floor without regarding the final status of ones they have previously issued. There could be many overlapping work orders on the floor in different states of completion. This leads to the waste of having excess inventories on the manufacturing floor.

In conclusion, work orders push materials to the manufacturing floor to meet the build plan called for by the MRP system and the master schedule. The lack of immediate feedback on the status of the materials previously released causes the floor to have too much inventory on hand in different states of completion. It also causes waste, since labor and overhead are used in tracking the work orders and the schedules.

2.2 PULL SYSTEMS

In a pull system, the consumption of materials rules the flow throughout the process rather than top-down schedules and releases. The last completion in the manufacturing process before the product arrives at the finished goods location is the pulling factor which moves the materials through the production line.

For example, imagine a long row of dominoes. In a push system, we would push the first domino of the row, this in turn would knock down the next one, and so on. Such a process pushes forward, and once it starts, it is very difficult to stop before the last piece is knocked down.

Suppose now that we connect the dominoes with an invisible thread so that there is plenty of slack and we would need to pull the pieces one by one in order to knock all of them down. Then we pull down the first domino, and after this one is down, we decide to pull over the one behind it, and so on. The result would be the same, but we have executed the task in a different way. The pieces are pulled down one by one.

The advantage of this system is that we can easily stop the dominoes from falling at any particular time. If one of the pieces has a problem and we want to stop the remaining pieces, we can stop pulling next piece in line and halt the falling process until the problem is solved.

Consider now the movement of materials in a factory during production. The production line will deal only with the actual materials needed to meet the build schedule of the product. Also, the process will stop itself quickly when a worker discovers a problem, and there will be no more consumption at the troubled work center. This operating system produces less excess material on the production line than in a push system.

To implement a pull system, the invisible thread that will pull down the domino needs to be created. Toyota solved this problem by implementing a Kanban system. (This system is the subject of Chapter 4.) But a Kanban system is not by itself a complete solution. A company also has to handle a cumbersome work order system in order to manage the build schedule and labor tracking. This task can be simplified by using a manufacturing system called *repetitive manufacturing*. (This system will be studied in the next chapter.)

A pull system has only one simple rule: *Move materials in the production line only when they are needed*. This means that we move materials by demand. In contrast, a push system moves materials by supply. Just-In-Time also calls for moving materials from one work center to another in the smallest possible quantities. The concept of the smallest quantity of material needed is a critical one for Just-In-Time systems. Conversely, it could be said that in a Just-In-Time system parts are never supplied to a process unless there is a demand for them. Excess building of parts is considered waste.

To further clarify this concept, we can use a simple example. Suppose we are building 50 units a day of a particular product. Also suppose that we cut a work order for the week and release a set of 250 parts to the manufacturing floor on Monday. According to the build schedule, we only need a set of 50 parts on Monday, but we have released 250, producing an excess of 200 sets that day, an excess of 150 sets on Tuesday, and so on. Only on Friday do we have the exact number of parts we need. In a Just-In-Time system, these excess sets would be considered a waste, for they aren't needed to build the daily quota. In such a system, we would only release 50 sets each day. If we wanted to fine-tune the system even further, we could release 25 sets every four hours. For high-volume applications, the frequency of releases could be increased (e.g., hourly schedules).

Figure 2.2 shows the flow of a pull system. Notice that the system uses the master

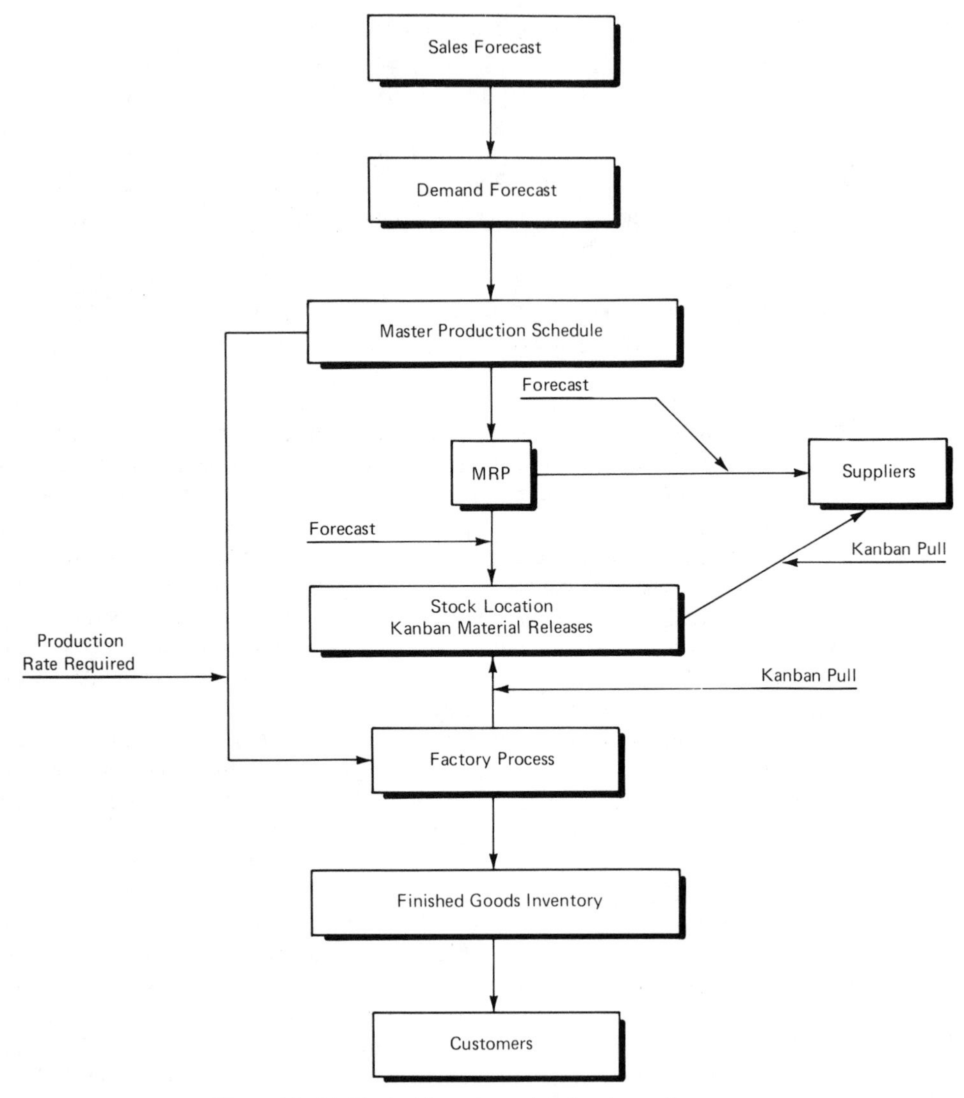

Figure 2.2 Build schedule and materials flow in a pull system.

schedule and MRP output to forecast the parts needed from suppliers and to establish the production rate schedules in the factory. Then, starting at the last process work center before the finished goods inventory, the system pulls the material required to build the products listed in the forecast. According to the figure, a Kanban system is used to pull the materials from the supplier center and from the stock location. The figure shows the pull process with a reverse arrow pointing up from the factory process to the stock location and from the stock location to the suppliers.

Materials Scheduling under a Pull System

One of the problems with a pull system is the lack of visibility for high-level planning. The problem arises in the translation of the sales forecast into materials planning and the detailed explosion of part requirements. In a push system, the production master schedule and the MRP system do the planning and the materials requirement explosion. This software also provides an inventory control system to track parts in inventory locations and all materials transactions in the factory. This part of the push system works well, but the system gets in trouble when planners cut purchase orders to suppliers and issue work orders to the factory without considering the actual materials needed on the manufacturing floor. Pull systems, conversely, work very well in self-adjusting the process variations once the factory's production rate is determined. The individual work centers don't need to know the complete picture to do a good job in scheduling materials through the production process.

By combining the best features of both systems, a company can implement an effective Just-In-Time system that will plan, forecast, and control the materials requirements in the factory. The production master schedule and the MRP system are used to do the high-level translation of the sales forecast into a production schedule and materials requirements to meet the build schedule. This information is used to alert the suppliers and the factory of the daily parts requirements. Then based on actual demand, the pull system moves the materials on the production line and from the suppliers.

Figure 2.3 shows the way in which a production master schedule and MRP system will accommodate a Kanban pull system. The idea is to use the MRP to plan the materials requirements and to use this information as a forecast to be provided to the different supply centers. Then, using a Kanban system, the materials are pulled from the suppliers at the time they are actually needed. This activity must be carefully coordinated so as to produce a realistic MRP output, for suppliers will use this information to plan the output of their production lines.

Figure 2.3 shows that the MRP system can be used to plan and procure parts handled in bulk issues. These parts, class C parts in inventory classification by volume and dollar value, are issued to the manufacturing floor in large quantities without regard to any build schedule. The C parts normally represent a lower dollar value in comparison with class A and B parts, which account for high volume and high value. Bulk-issue parts are normally expensed at the time the stockroom releases them to the manufacturing floor.

Just-In-Time calls for a selected set of quality suppliers that commit to frequent deliveries per material demand requests. One way to reward suppliers for this extra service

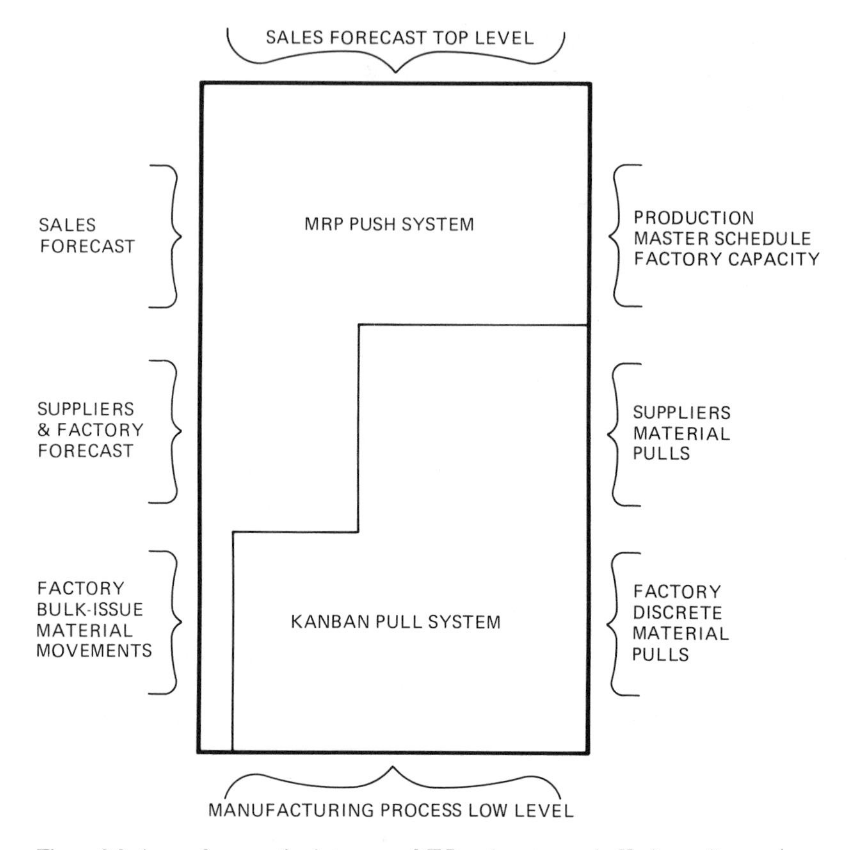

Figure 2.3 Areas of cooperation between an MRP push system and a Kanban pull system.

is to offer them a single-source, long-term commitment. The MRP system can produce a forecast to provide to the suppliers and a Kanban system can pull materials as they are required. Chapters 5 and 6 describe how to arrange this operating mode.

Figure 2.4 shows a supply-demand forecast organized by part numbers and by requirements per month. Notice that in this forecast there is a line ("Ship") that forecasts the number of units the sales force will sell during a particular month. The "Build" line represents the commitment of the manufacturing organization to manufacture the number of units required to support the "Ship" forecast. The "F/G" entry gives the number of units planned to be stocked in a particular month to support unforecasted sales. The sales force must be responsible for the products cited in this entry.

Figure 2.5 shows the output of the Ask MANMAN MRP system for a single part number. The basic supply-demand forecast provides the input to the MRP in the form of a master schedule. The MRP then prints out a report for each part number showing the availability of parts to meet the schedule. The MRP report will also show the order points for additional parts based on the procurement lead times entered in the computer.

Supply/Demand Forecast FINAL

Model/Feature		JUNE	JULY	AUG	SEPT	OCT	NOV	DEC
1200	New Sales	552	542	562	582	552	562	572
98-1008-00	Add-Ons	30	40	40	40	40	50	60
(97-1011-00)	In-House	0	0	0	0	0	0	0
Workstation II	WS II Swaps	20	20	20	20	20	20	20
Mo. Avg = 575	Ship	602	602	622	642	612	632	652
	Build	600	600	600	650	600	650	650
	Cum Demand	5774	6374	6974	7624	8224	8874	9524
	Cum Supply	5776	6376	6976	7626	8226	8876	9526
	Cum Variance	2	2	2	2	2	2	2
	F/G	110	108	86	94	82	100	98
121	New Sales	414	407	422	437	414	422	429
98-0101-00	Add-Ons	23	30	30	30	30	38	45
(97-0105-00)	In-House	0	0	0	0	0	0	0
Numeric Keypad	WS II Swaps	15	15	15	15	15	15	15
Mo. Avg = 440	Ship	452	452	467	482	459	475	489
	Build	450	450	450	488	450	488	488
	Cum Demand	5070	5520	5970	6458	6908	7396	7884
	Cum Supply	5070	5520	5970	6458	6908	7396	7884
	Cum Variance	0	0	0	0	0	0	0
	F/G	106	104	87	93	84	97	96
2101	New Sales	75	74	76	78	75	76	80
97-1020-00	Add-Ons	5	5	5	5	5	5	7
File Server Frame	In-House	0	0	0	0	0	0	0
Mo Avg = 78	Ship	80	79	81	83	80	81	87
	Build FS 2	80	80	80	80	80	81	87
	Cum Demand	149	229	309	389	469	550	637
	Cum Supply	149	229	309	389	469	550	637
	Cum Variance	0	0	0	0	0	0	0
	F/G 1	1	1	1	1	1	1	1
	F/G 2	4	5	4	1	1	1	1
215	New Sales	1	1	1	1	1	1	1
97-1026-00	Add-Ons	0	0	0	0	0	0	0
FS II Upgrade Kit	In-House	0	0	0	0	0	0	0

Figure 2.4 Sample of a production master schedule.

```
RE,907  .MDATABAS   METAPHOR COMPUTER SYSTEMS
THU, MAR 17, 1988,   2:53 PM    MRP RUN:  3/16/88     REQUIREMENTS REPORT BY PART                                               PAGE NO:   2

PART NUMBER          RV DESCRIPTION                         UM SC CCODE ABC OPC    EOQ      SAFETY  MULTIPLE SHRINK  N DAYS          BC
------------------   -- ------------------------------      -- -- ----- --- ---  --------  -------- -------- ------  --------        --
10-0103-01           A  ANTI-GLARE SCREEN                   EA BE    7   B   1        1        50        0    0.0        0          JB

   QOH      UNIT CST     VALUE    INLOC       OPEN ORD GROSS RQ  FLTIME    ULTIME  DKTOSTCK   VENDOR
--------  --------  ----------  ----------  --------  --------  --------  --------  --------  ----------
     162     15.65    2534.90 MS                 400      1163        60    0.0000         5 02680

                                            DUE                     DEL                                                  DUE                 DEL
*  ORDER NO  QUANTITY  VALUE      VENDOR     DATE  MRP NEED STA DATE  *  ORDER NO  QUANTITY  VALUE     VENDOR     DATE  MRP NEED STA DATE
- ---------- -------- ------- ----------   ----- -------- --- ----- - ---------- -------- ------- ----------  ----- -------- --- -----
  PR712061        100 1564.76 002680      3/21  4/14/88       4/07!  PR712061        100 1564.76 002680     4/11  6/02/88       5/25
  PR712061        100 1564.76 002680      5/09  7/07/88       6/29!  PR712061        100 1564.76 002680     6/01  8/04/88       7/28
```

	PAST DUE	3/14/88	3/21/88	3/28/88	4/04/88	4/11/88	4/18/88	4/25/88	5/02/88	5/09/88	5/16/88	5/23/88	5/30/88
GROSS REQTS:	0	41	0	38	0	42	0	0	38	0	38	0	38
OPEN ORDER :	0	0	100	0	0	100	0	0	0	100	0	0	100
ORDER DUE :	0	0	0	0	0	0	0	0	0	0	0	0	0
ORDER START:	0	0	0	0	0	0	0	0	0	0	0	0	0
PROJ AVAIL :	162	121	221	183	183	241	241	241	203	303	265	265	327
RESCHED TO :	0	0	100	100	100	100	100	100	100	200	200	200	200

	6/06/88	6/13/88	6/20/88	6/27/88	7/04/88	7/11/88	7/18/88	7/25/88	8/01/88	8/08/88	8/15/88	8/22/88	8/29/88
GROSS REQTS:	4	37	0	31	25	46	25	0	32	31	32	31	34
OPEN ORDER :	0	0	0	0	0	0	0	0	0	0	0	0	0
ORDER DUE :	0	0	0	0	0	0	0	0	0	0	0	0	0
ORDER START:	0	0	0	0	0	0	0	0	0	0	0	0	0
PROJ AVAIL :	323	286	286	255	230	184	159	159	127	96	64	33	-1
RESCHED TO :	200	200	200	200	100	100	100	100	0	0	0	0	0

	9/05/88	9/12/88	9/19/88	9/26/88	10/03/88	10/10/88	10/17/88	10/24/88	10/31/88	11/07/88	11/14/88	11/21/88	FUTURE
GROSS REQTS:	34	38	34	0	29	29	29	29	32	31	32	31	252
OPEN ORDER :	0	0	0	0	0	0	0	0	0	0	0	0	0
ORDER DUE :	0	0	0	0	0	0	0	0	0	0	0	0	0
ORDER START:	0	0	0	0	0	0	0	0	0	0	0	0	0
PROJ AVAIL :	-35	-73	-107	-107	-136	-165	-194	-223	-255	-286	-318	-349	-601
RESCHED TO :	0	0	0	0	0	0	0	0	0	0	0	0	0

WHERE REQUIRED	DEMAND TY	QUANT	DATE	WHERE REQUIRED	DEMAND TY	QUANT	DATE	WHERE REQUIRED	DEMAND TY	QUANT	DATE
97-1011-02	FORECAST	31	3/17	96-1051-03	BLD SCHED	10	3/17	97-1011-02	FORECAST	38	3/31
97-1011-02	FORECAST	37	4/14	96-1051-03	FORECAST	5	4/15	97-1011-02	FORECAST	38	5/05
97-1011-02	FORECAST	38	5/19	97-1011-02	FORECAST	38	6/02	ORDER DEMAND FILE	FORECAST	4	6/10
97-1011-02	FORECAST	37	6/16	97-1011-02	FORECAST	31	6/30	97-1011-02	FORECAST	25	7/07
97-1011-02	FORECAST	31	7/14	96-1051-03	FORECAST	15	7/15	97-1011-02	FORECAST	25	7/21
97-1011-02	FORECAST	32	8/04	97-1011-02	FORECAST	31	8/11	97-1011-02	FORECAST	32	8/18
97-1011-02	FORECAST	31	8/25	97-1011-02	FORECAST	34	9/01	97-1011-02	FORECAST	34	9/08
97-1011-02	FORECAST	34	9/15	ORDER DEMAND FILE	FORECAST	4	9/16	97-1011-02	FORECAST	34	9/22
97-1011-02	FORECAST	29	10/03	97-1011-02	FORECAST	29	10/13	97-1011-02	FORECAST	29	10/20
97-1011-02	FORECAST	29	10/27	97-1011-02	FORECAST	32	11/03	97-1011-02	FORECAST	31	11/10
97-1011-02	FORECAST	32	11/17	97-1011-02	FORECAST	31	11/22	97-1011-02	FORECAST	32	12/08
97-1011-02	FORECAST	31	12/15	97-1011-02	FORECAST	31	12/22	97-1011-02	FORECAST	32	12/22
97-1011-02	FORECAST	32	1/05	97-1011-02	FORECAST	31	1/12	97-1011-02	FORECAST	32	1/19

ADDITIONAL REQUIREMENTS ON FILE   RESCHEDULED   PN:10-0103-01

Figure 2.5 Sample of an MRP output.

2.3 MATERIALS MOVEMENT IN THE FACTORY

There are two types of materials movement in a manufacturing environment. The first is the external movement of materials from the suppliers to the factory. The second one is the internal movement of materials through the production process.

It's fairly obvious that deliveries by suppliers are the hardest to control. In this case, a company is dealing with other companies, including probably several that do not have a Just-In-Time system and instead operate in batch modes. Also, there could well be transportation problems, especially if materials have to travel from one side of the country to the other or, even worse, across the ocean from another country. In a Just-In-Time system, a company spends considerable time coordinating and improving both types of movement. The time that the materials spend traveling does not add any value to the final product and is therefore a waste. Chapters 6 and 7 deal with the selection of suppliers and the advantage of dealing with suppliers close to the factory to minimize transportation waste.

Supplier Deliveries

The criteria for orchestrating the suppliers' deliveries are dependent on many factors related to the type of product, volume, cost, and supplier. These factors determine the policy to use with a given supplier. For example, a company wants to work closely with those suppliers that ship the highest dollar-related parts. This has to do not only with the price of the parts but also with the volume. The volume usage of the part and its price relate to the dollar ratio of the part. The company might use a very cheap part—let's say it costs a dollar—but it uses one million of them per day. It should pay much more attention to this part than a part that costs a hundred thousand dollars but is used at a rate of one per month.

Depending on the volume and cost ratios, a company might use a Kanban system to control the delivery of the parts. The basic principle is to ask the supplier to deliver frequently and in the smallest lots that will meet daily production needs. For high-volume operations, the company could have a group of suppliers delivering parts several times a day at specific times. The principle here is to get frequent deliveries of the smallest lot sizes and that occur only when the production line pulls the material.

One simple way to ensure the frequent delivery of parts is to set up a truck route and have a truck pick up and deliver parts from several suppliers at the same time. The manufacturing company could plan this truck route in such a way that the truck picks up the parts at the suppliers' shipping docks according to a prearranged schedule. Any truck company will provide this service, or if the suppliers are local, the manufacturing company could use its own truck and driver.

The most common objection from suppliers to operating in this mode is that it requires them to synchronize their factories to the factories they ship to. There are several ways to solve this problem, and they will be described in later chapters. The other objection to frequent deliveries is that they increase the overhead associated with receiving and inspecting deliveries. The solution to this problem will be discussed in Chapter 8.

An increase in invoicing paperwork and in the workload of the accounts payable department is another problem caused by frequent deliveries. Chapter 9 includes a discussion of the solution to this problem.

Factory Materials Movement

The movement of materials in the factory offers different problems from those associated with the external movement of materials. In the factory, the company has direct control and can arrange the system to operate using Just-In-Time principles. This task, however, is not simple, for problems arise that involve people. Workers and supervisors tend to believe that working with small lots of materials means more work for them, although actually the system becomes more repetitive and easier to handle than with larger lots.

The key to a successful repetitive lot system is to simplify the material tracking system so that the overhead associated with the handling of many small lots doesn't become excessive. This is the place where the concepts of repetitive manufacturing and Kanban play an important role in the simplification of the production process.

A work order system would be difficult to implement in a small-lot system because of the overhead associated with the opening, tracking, and closing work orders; the number of work orders would be too large and the system would be ineffective.

Daily Materials Issuing

One of the fundamental changes that occur in a Just-In-Time system is the conversion from a monthly production schedule to a daily one. The factory is converted into an operation that accounts for daily completions. Assume the company is building two thousand units a month of a particular product and is working with work orders of five hundred units at a time. Then it converts the schedule to a daily completion quota of one hundred units. This quota should be made known to all the workers in the factory and to the staff of the materials department. The materials department could then issue materials to the factory floor two to four times per shift in quantities of twenty-five to fifty.

Implementing daily schedules is usually resisted in cases where the product is not volume-oriented. For example, suppose a company is building large mainframes at a production rate of six per month. The solution is to identify subassemblies of the mainframe that could be scheduled as daily builds. Then, a final assembly operation at the end of the production line would take care of the six completions.

Note that the mainframe's subassemblies don't necessarily have to be of the same type. For example, the company could build memory modules for a couple of days. The next day it could build I/O modules and the next day processors. The point is that the company could always find some type of manufacturing activity that could be scheduled in daily rates. The principle applies even if the company doesn't have enough volume to fill a full month's schedule with the same product. Daily schedules allow the company to measure production results in small increments—a critical concept of Just-In-Time.

2.4 ELIMINATE SAFETY STOCKS

Just-In-Time considers safety stocks a major source of waste, for they don't serve any useful purpose other than to cover for deficiencies in manufacturing. By nature, people want to feel safe in everything they do. Having safety stocks is a simple way to feel secure in a manufacturing environment. But safety stocks are expensive and wasteful and could create many control and obsolescence problems. Of course, safety stocks are very useful if a company wants to operate in a Just-In-Case mode, but this mode is unpredictable and conflicts with Just-In-Time.

There are two types of safety stocks. The first type is voluntary. The materials organization uses voluntary safety stocks to cover wrinkles in the planning process or to compensate for a supplier's poor performance. These are the safety stocks that the materials staff can easily talk themselves into implementing.

The second type is involuntary. Involuntary safety stocks occur when the materials department issues more parts to the manufacturing floor than the production process can handle. For example, assume the department issues a lot of one hundred parts to the production line, but the process capacity is only ten per day. This means that the factory is going to have an excess of ninety parts in the WIP inventory the first day. Then it will have an excess of eighty parts on the second day, and so forth. This excess WIP inventory is like a disease that should be cured, for it slows down the timely solution of problems in the production process rather than bringing them into the open as soon as they occur.

To see why this is so, imagine a worker building ten units per day but with two hundred sets of parts at hand. When a part is defective, the worker most likely is going to set it aside and borrow from the other one hundred ninety to complete his daily quota. The company measures workers by completions, so the worker's first priority will be to complete the ten units. The next day, he borrows again from the remaining excess to complete the next ten units. At the end of the month, his stock is down to the defective parts he put aside. The result is that whatever problem the defective parts had is not dealt with until the end of the month. By then, another two hundred parts will have been released to the floor, some of which will probably need urgent repair.

Under a Just-In-Time system, the worker has only ten sets of parts to build the ten units for the day. If some of those units are defective, he will not be able to complete his daily goal, and the supervisor will be forced to address the problem immediately.

Line Balancing and Buffers on the Production Line

Buffer inventories placed along the production floor also create the problem of hiding imbalances in the process. Proper line balancing in a process is critical for obtaining high worker efficiency. But when there are buffers between work centers, any differences in the balance of the line are less evident. Workers are operating from and to pools of parts that mask any output differences. This causes the problem that some workers have to try very hard to catch up while others have to slow down to keep pace with the process flow. Just-In-Time considers such slowing down to be a waste.

The ideal buffer inventory between two work centers consists of one part. When a worker finishes an assembly, another one is available immediately. Having additional assemblies available would be a waste not only of materials but also of the labor invested in the parts in the buffer.

2.5 SUMMARY

The key to the movement of materials is to move them frequently, in small lots, and only when they are needed. This approach is consistent with the use of a repetitive lot system and a pull system. To be successful, a Just-In-Time system requires much time spent reducing the lot sizes of materials and increasing the frequency of movement. This applies to both the factory and the suppliers.

It is recommended not to go overboard by implementing the concept with respect to all the parts required for production. It is best to develop a plan in which the parts are ranked. Then start to implement the system in sequential steps. First concentrate on the highest-dollar related parts and the most important suppliers. Once those are taken care of, go to the next level of parts.

One word of caution: Do not implement a system of frequent deliveries without simplifying the paperwork, including invoicing and accounts payable. Also, institute a system to control the movement of materials from one place to another. A Kanban system might be the solution, although there are other ways to obtain the same results. It is possible to be creative in the operating mode as long as the new system results in small daily deliveries with no buffers to compensate for problems and line imbalances. Chapter 9 describes some ways to simplify the paperwork in materials and accounting departments.

It is important not to reduce buffer inventories and go to small deliveries until the quality of parts reaches an acceptable level. Not including quality in the equation would be a sure way to bring the production line to a screeching halt. Start solving quality problems with suppliers and the production process long before you start reducing inventory.

You may consider using a reduction in inventories to a lower level as a way to force the solution of quality problems. This approach, if not executed carefully, will impact the shipments out of the factory. Management must understand that the poor results are not due to the Just-In-Time system. Be wary of painting yourself into a corner.

REFERENCES

BELT, BILL, "MRP and Kanban: A Possible Synergy?" *Production and Inventory Management* (American Production and Inventory Control Society) 28, no. 1 (1987): 71–80.

GROUT, JOHN R., and MARK E. SEASTRAND, "Multiple Operation Lot Sizing in a Just-In-Time Environment." *Production and Inventory Management* (American Production and Inventory Control Society) 28, no. 1 (1987): 23–27.

HALL, ROBERT W. *Zero Inventories*. Homewood, Ill.: Dow Jones-Irwin, 1983.

JAPAN MANAGEMENT ASSOCIATION, ed. *Kanban: Just-In-Time at Toyota.* Translated by David J. Lu. Stamford, Conn.: Productivity Press, 1985.

SCHONBERGER, RICHARD J. *Japanese Manufacturing Techniques: Nine Hidden Lessons in Simplicity.* New York: The Free Press, 1982.

SUZAKI, KIYOSHI. "Corporate Culture for JIT." Paper presented at APICS Zero Inventory Philosophy and Practice Seminar, St. Louis, 1984.

———. "Work-in-Process Management: An Illustrated Guide to Productivity Improvement." *Production and Inventory Management* (American Production and Inventory Control Society), Fall 1985, 101–110.

3 ★

A Lean Factory Under a Just-In-Time System

Just-In-Time places a high priority on reducing idle materials on the manufacturing floor. Not only are such materials, in the form of buffer inventories, a waste from the point of view of cash investment, but they also hide latent problems in quality and in the manufacturing process. Idle materials are easy to discover on the manufacturing floor, but their causes are sometimes not so evident. Possible causes include line imbalances in the production process, yield problems in a particular work center, or even procedural problems that result in the issuing of unneeded materials. Just-In-Time offers a systematic way to analyze material excesses and provide solutions to eliminate them.

This chapter reviews the basic set of rules that Just-In-Time uses to issue materials to the manufacturing floor. These rules minimize the use of buffer inventories, and they will help to simplify the process of pulling and issuing the parts in the stockroom. Also reviewed are the concepts and procedures required to implement a repetitive manufacturing system to eliminate work orders and related paperwork.

Manufacturing organizations must have a clear procedure to issue materials to the production floor. This procedure has to deal with three critical aspects of the job. First, the procedure must ensure the materials issued are sufficient for the production build schedule. This involves determining hourly, daily, or weekly schedules as well as material lot sizes and stocking locations. Second, the procedure must allow the company to track the materials moving through the production process so that at any particular time it can discover exactly which materials are in which locations and how much labor has already been applied to them. Third, the procedure must allow analysis of the physical movement of materials in the factory so as to be able to increase productivity and to reduce overhead.

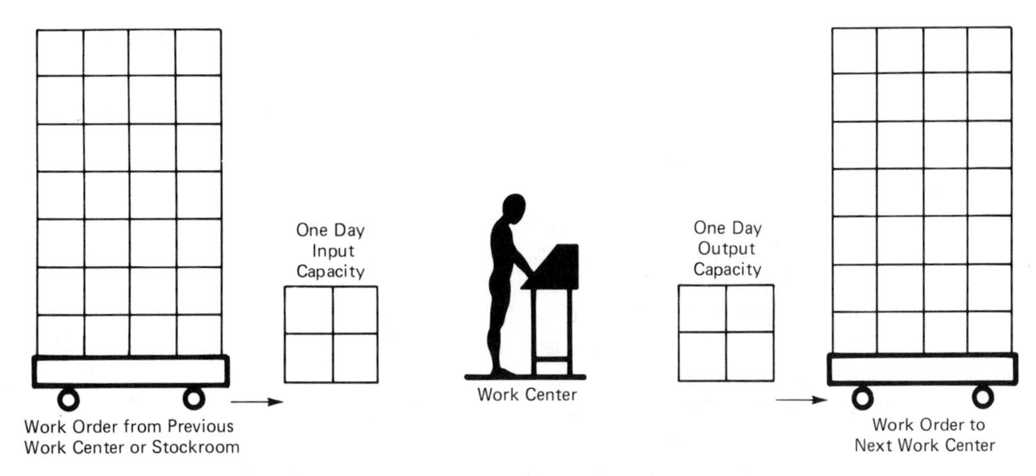

Figure 3.1 Batch-mode work order flow of materials.

Just-In-Time takes a hard look at issuing, tracking, and moving materials in order to minimize the amount of idle materials at work centers during production. Just-In-Time also looks at buffer inventories between work centers, direct labor waste, paperwork to track materials, and the waste of labor to move materials around. This chapter and the next deal with issuing and tracking materials. Chapter 5 deals with the movement of materials and time waste.

Figure 3.1 shows the materials flow in a batch mode, work order system. Notice that the materials move in quantities larger than the number of units that a worker can process in a single work shift. These excess materials result in idle buffers between work centers that waste the money invested in them. This mode of operation makes it easier to borrow parts from the buffers to cover for defective parts, causing a delay in reporting quality or design problems with the product.

Figure 3.2 shows a continuous flow of materials, with buffers between work centers reduced to the number of units that a worker can process in a work shift. Any problems with

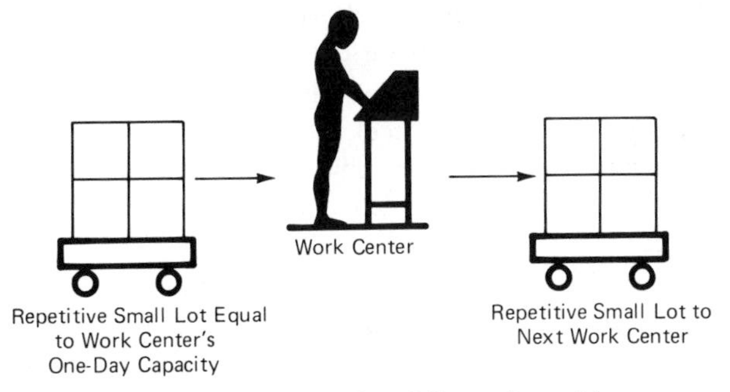

Figure 3.2 Continuous flow delivery of materials.

parts in a work center will be immediately evident, because the worker will not be able to complete the assigned task.

3.1 MATERIALS DELIVERY TO WORK CENTERS

Delivering materials to work centers is a continuous process that must match the factory flow with the exact type and number of parts required. This job is normally labor intensive and very critical in sustaining the operation of the factory. The materials are the fuel for the manufacturing engine, for there would be no activity in the factory without the proper flow of parts. Just-In-Time provides a set of rules that regulates the issuing of materials to the manufacturing floor. These rules are simple, but they can have a devastating effect on production output if implemented without the support of Just-In-Time programs in other areas of the organization.

The rules for issuing materials to the production floor are as follows:

- Deliver materials to the production line on a daily basis and in the amount required for a day's production.
- If volume is very high, then deliver many times in a day. Never deliver more than one day's production.
- If the production line didn't finish building the scheduled quota, then the materials issued the next day should net the daily requirements. Never issue full requirement sets to a noncompleted production schedule.
- Never issue materials to the line with part shortages. A shortage will produce a line stop.
- Issue materials to the line on a rate basis. Use repetitive manufacturing software to execute a backflush at the end of the line in order to deplete the parts from the WIP location. Don't use work orders to control materials and the production flow.
- Absorb labor on the product's completion only. Don't subdivide the earning of labor into modules, which makes the system more complicated.
- Allow the storing of free stock parts on the production line for more than one day. Select a procedure to replenish these parts and don't track them by unit. One way to do this is to spend them as they are issued.
- Pull large parts from the stock location on a demand basis. One way to do this is to use an intercom system to allow workers to call for the parts just before they need them. Implement a simple delivery system that does not need the intervention of line workers.
- Use Go/No Go tests for the production line. Never troubleshoot during production. Replace defective parts for good ones and take the faulty parts off-line for repair. Treat all faulty parts with special attention. They should not have reached that point in the production process. Take corrective actions to solve the problem immediately.

- Give authority to the workers to stop the line whenever they discover a problem. This will provide the workers with a sense of ownership and they will be more likely to try to improve the system.

The first to raise objections to the above set of rules will be the materials department staff. They will worry about two issues. First, the rules require them to issue materials with no shortages. They will claim that in some cases the workers could save time by building partial assemblies. Then the workers could quickly complete them when the short parts arrive at the factory. This is an easy trap to fall into, but it should be avoided. To begin with, the time invested in building partial assemblies can be used to build other products. Furthermore, there is no bigger incentive for the materials department to solve a part shortage problem than to have a line down.

The main issue that arises when there is a shortage is not the procurement of parts to make up the shortage. Rather it is to understand why the parts were short. Knowing this allows a company to take corrective action to avoid a repetition of the problem. For example, problems with the planning process or the purchase of the parts might be the real reason for the shortage. There could also be a problem with the supplier or with the procedures to track and expedite late deliveries. Understanding and correcting the causes of a parts shortage is more important than procuring the parts.

The second objection to the above rules is that additional labor is required to pull smaller but more frequent lots of parts. The most obvious way to solve this problem is to allocate more people to the stockroom. Time, however, will prove that this approach is not necessary; after a while, the stockroom clerks will become more productive doing small repetitive jobs.

There are other ways to reduce the labor involved in pulling and releasing parts. One is to implement a dock-to-WIP system, which bypasses the stockroom entirely. With a dock-to-WIP system, a company moves parts from the receiving dock to the manufacturing line directly, without stopping in the receiving inspection department or the stockroom. This approach calls for parts of excellent quality from the supplier and for the delivery of the exact quantity of parts needed at the work centers. Dock-to-WIP systems are discussed in detail in later chapters.

Another way to reduce the additional labor required for frequent kit pulling is to ask the supplier to pack the parts in containers holding convenient amounts—this saves time in counting. Consideration needs to be given to the parts usage rate and the ease of handling the containers on the manufacturing floor.

3.2 REPETITIVE MANUFACTURING

Repetitive manufacturing is an alternative to a work order system for production scheduling, material issuing, and tracking of parts and labor throughout production. Repetitive manufacturing uses continuous production rates rather than discrete, batch-oriented schedules.

Work Order Systems

Before discussing repetitive manufacturing, let us consider work order systems. A work order system is a batch mode process in which the materials department releases materials to the manufacturing floor in discrete quantities called work orders. Department staff normally open work orders on a computer and assign the job a kit quantity with beginning and completion dates. Then they send the paperwork to the stockroom for execution.

Work order releases usually cover several days or weeks of production. The time span of a work order will depend on the production rates and the policies effective in the materials department.

A work order system binds all the parts released with the work order to the work order's paperwork. The parts and the papers must travel throughout the production line together.

Work orders are also financial entities from the cost point of view. The computer will assign them a standard labor and materials cost. During the execution of a work order, the company tracks labor and adds its value to the cost of the materials. The computer then compares the actual costs against the standard costs associated with the work order. This comparison will determine the efficiency of the manufacturing process in building the work order.

Let's consider a simple example of a work order release. Assume that a company is building 500 units per month of a particular product. Using a work order system, the company could cut two work orders per month of 250 parts each, assigning a life span to the work orders of two weeks. It is clear that the company is releasing more material to the floor than is actually needed for one day of production.

The problem with work orders is that they are difficult to track. For example, the dates framing the opening and closing of the work orders are more flexible than with a repetitive system. A company is working with time spans of weeks rather than a few hours or a day. The possible overlap of two work orders for the same job causes the problem that materials from one work order might be borrowed for the other, making it impossible to close the work order that is incomplete. Finally, because of the large size of the work orders, there is sometimes a need to split them at the work center to move partial completions on the line. Doing this increases the overhead required to maintain the open work orders on the computer and on the manufacturing floor. This overhead constitutes waste in a Just-In-Time system.

Repetitive Manufacturing Systems

Repetitive manufacturing is a production system in which a company issues parts to a production floor and builds products on a rate basis. No work orders are required to issue parts and to build products. The only thing needed is to translate the production master schedule into a build rate and then consistently issue the parts at that rate.

The most efficient way to handle repetitive manufacturing is to set up a daily rate. Then a computer system with a repetitive manufacturing software module can issue all the parts that are needed to build at that rate. These parts, once released to the manufacturing floor, are not allocated to any particular paperwork. They are part of a pool of parts that form the WIP.

If a company is building different products with common parts and at common work centers, the repetitive software on the computer would net the requirements and issue the common parts at the time they are needed at the work centers. This is unlike a work order system, where all the parts associated with a work order are issued in a specific batch. A company never mixes parts of different work orders. The repetitive netting process simplifies the additional counting and traveling to the stockroom to pull the parts.

Backflushing Completions

A repetitive manufacturing system closes the completion of an assembly with a backflush transaction. Backflushing an assembly is a reverse operation on the computer. The system explodes the assembly's bill of materials backwards, depleting all the parts from the WIP location where the assembly was built. The computer replaces these parts with a single higher-assembly number.

A backflush transaction in a work center decreases the value of the WIP location by the number of components specified on the bill of materials as having been used to build the assembly. Backflush also credits the labor earned in building the assembly and the overhead associated with it. In essence, a backflush transaction does the tracking work performed in a work order system, but without the associated paperwork. Backflushes are simpler operations for small, frequent parts completions.

When a company establishes the process routing, it must determine carefully at which points to execute the backflush transactions. The manufacturing engineers, in agreement with the materials and cost accounting staff, will select the right places to set up the backflushing gates. The placement of the gates is influenced by the structure of the bill of materials, the nature of the process, and the physical layout of the work centers on the manufacturing floor.

It is recommended that manufacturing set up as few backflushing gates as possible to minimize the number of transactions and process fences (i.e., physical boundaries between different operations in the process). This is sometimes not easy to do, because of the structure of the product's bill of materials. But even if a company has to change the bill of materials to reduce the number of backflushing points, it would pay in the long run as a result of the savings in the number of transactions.

Figure 3.3 shows the location of the backflushing gates in a production process to build a final product C. The assumption is that processes A and B are located in different places than the final process C. In this case, there is no other choice but to backflush at the end of processes A and B.

Notice that all the parts for processes A, B, and C are issued, in a repetitive mode, from the same stock location. Also notice that the issuing of parts C is offset so they arrive at process C at the same time as subassemblies A2 and B3. The repetitive software that issues parts will keep track of the offset of parts C with respect to parts A and B. Offsetting will be explained in more detail in the next section.

Figure 3.4 shows the same process, but it represents the case where processes A and B are at the same location as process C. In this case, there is no need to backflush at the end

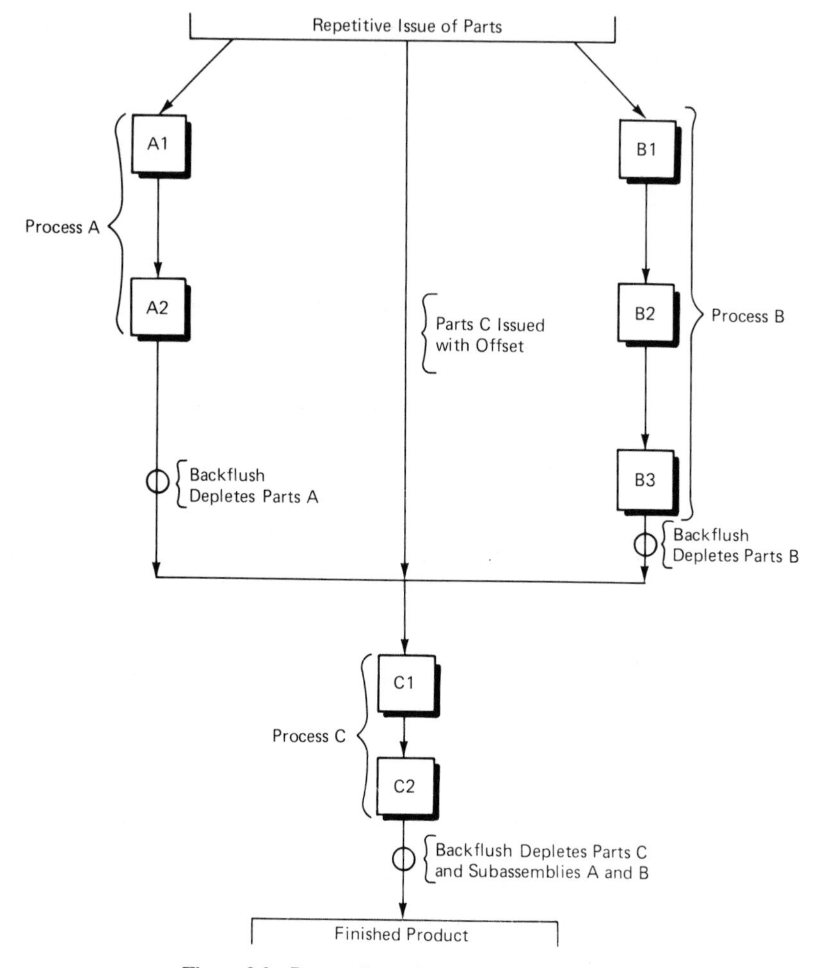

Figure 3.3 Process flow with three backflush gates.

of processes A and B. A single backflush at the end of process C will deplete all the parts for the three processes.

3.3 ASK MANMAN/REPETITIVE

To further explain the concept of repetitive manufacturing, this section contains an overview of the Ask MANMAN/REPETITIVE system, an actual software package.

Ask Computer Systems was one of the first software houses that released a package to handle manufacturing in a repetitive environment. Like other such software, the Ask pack-

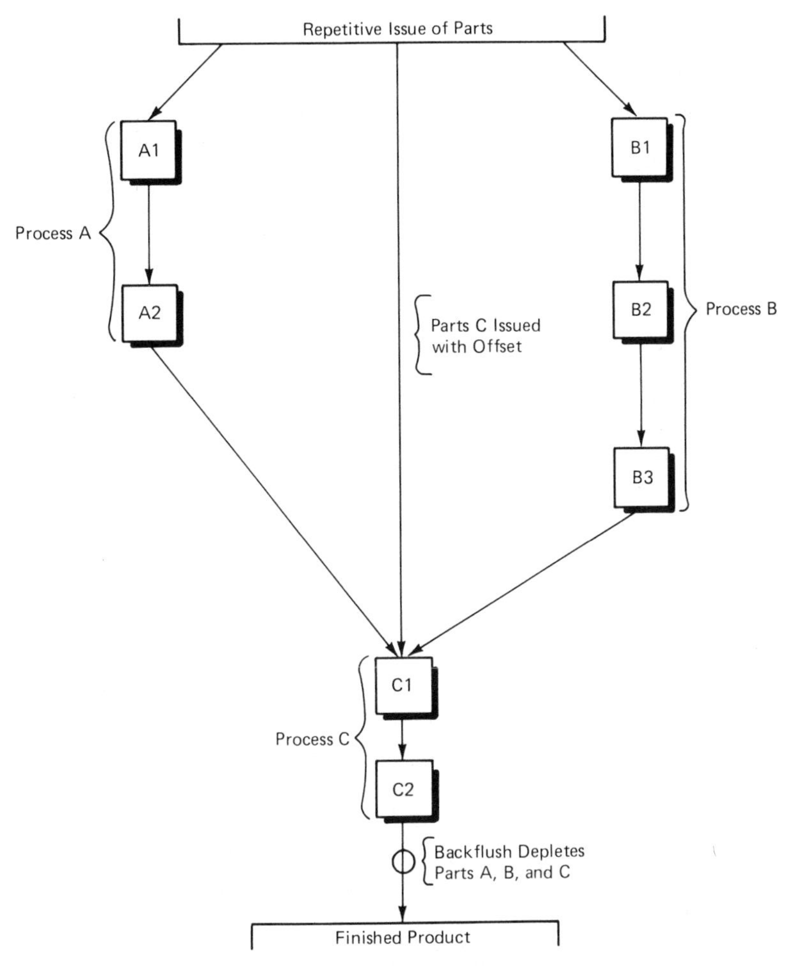

Figure 3.4 Process flow with one backflush gate.

age was initially a modified version of a work order system changed to handle production rates instead of discrete work orders. After a couple of releases, Ask improved this package and offered the many repetitive features necessary for this type of process. The Ask package runs in a Hewlett Packard HP3000 computer, and the package works in coordination with the Ask MANMAN/MFG system.

The MANMAN/REPETITIVE Process

The MANMAN/REPETITIVE process consists of four operational steps. Three departments in the manufacturing organization execute these steps on a repetitive basis.

First, the production control group creates and maintains the repetitive schedule based on the sales forecast and the build plan. The build plan is normally updated once a month after a new sales forecast.

The stockroom clerks execute two transactions. First, they run a pull list report that lists all the parts required to build the product in a repetitive fashion. Then, after they pull all the parts, the clerks send the parts to the corresponding work centers.

Finally, after the corresponding work center has assembled the parts into higher assemblies or final products, a worker runs a backflush transaction at the end of the production line. This transaction credits all the parts used and replaces the inventory record by a single higher assembly number.

Figure 3.5 shows the flow diagram for the transactions mentioned above. It also shows the corresponding transaction number in the MANMAN/REPETITIVE system. The events that occur in every one of these transactions are described in detail later in the chapter.

Initial MANMAN/REPETITIVE System Setup

Before starting to operate the factory in a repetitive mode, a series of data definition records must be entered into the computer to define the process and the product structure. The system requires that these records be entered using the MANMAN/MFG system. MANMAN/REPETITIVE uses this information later as part of the normal repetitive operation in the factory.

There are four sets of records required to define the process and the product: manufacturing account records, WIP inventory records, part records, and bill of material records.

Manufacturing Account Records

Two manufacturing account records must be set up to collect material, labor, variances, and WIP inventory values. The system associates these accounts with locations on the manufacturing floor.

Work area clearing account. This account collects the material and labor cost variances of the work centers. Also, if someone backflushes from an intermediate count point, the account will collect the WIP inventory values.

WIP location accounts. These accounts collect the value of WIP inventory on the manufacturing floor. They directly relate to a physical location on the floor.

WIP Inventory Location Records

MANMAN/REPETITIVE uses two types of inventory locations: (1) work center WIP locations (where parts are processed) and (2) WIP inventory locations (where parts are stored for the process).

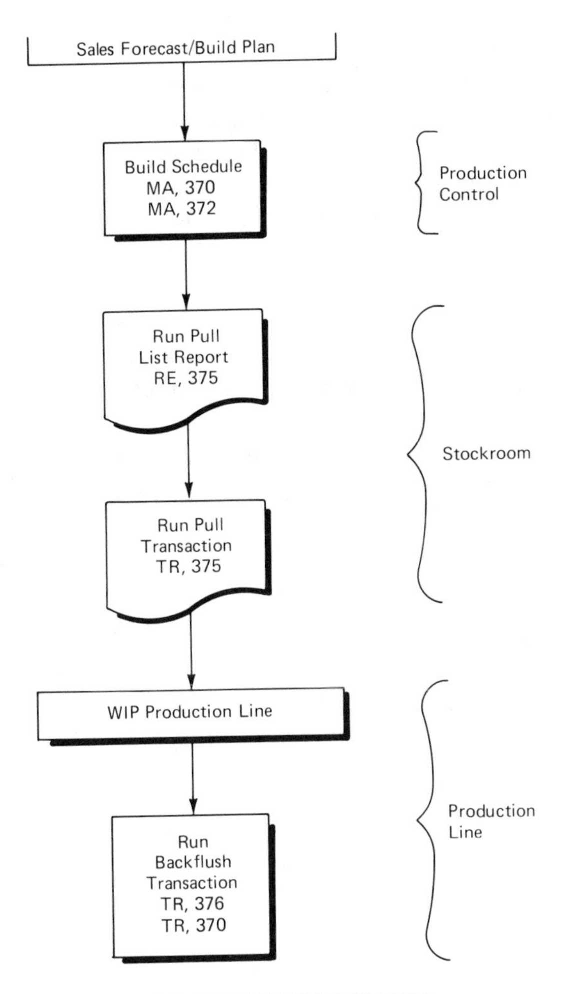

Figure 3.5 MANMAN/REPETITIVE Process.

Work center WIP locations. A WIP location number must be assigned to each work center defined in the repetitive process. There is no limit to the number of process steps that a work center can include.

There could be a single work center WIP location for the whole process, which would minimize the number of transactions in the system. Together Figures 3.3 and 3.4 show how this would work. The process in Figure 3.3 requires the definition of three work center WIP locations and three backflush gates. To reduce number of transactions, a manufacturer could use the process shown in Figure 3.4, which has only one work center WIP location and a single backflush. In a Just-In-Time system, the process in Figure 3.4 is ideal, for it reduces to a minimum the number of transactions required for operation.

WIP inventory locations. These locations are the areas where the repetitive process will send the repetitive assemblies for storage. A company must define the WIP locations in the system before it can store or pull parts from those locations. Just-In-Time considers these locations a waste and would have them reduced or eliminated.

Parts Issuing Policies

A manufacturer must define on the computer the issue policies for every part in the system. These issue modifiers allow the manufacturer to define the number of days' supply, the pansize, and the safety stocks for each part individually. In a Just-In-Time system, the only parts issued to the floor are the ones needed for the day's production.

Number of days supply. This modifier allows a manufacturer to issue the repetitive parts for more than one day's supply. This feature could be used in a repetitive system to issue small parts that don't need to be counted daily. These parts fall into the category of bulk-issue parts, which might be handled weekly or at longer time intervals.

Pansize. Another factor to consider when pulling parts from a stock location is the containers in which the parts are packed. This modifier allows a manufacturer to specify a quantity for issuing the parts. For example, if the supplier had packed the parts in boxes of 12 and the manufacturer requires a daily pull of 100, the repetitive system will issue a quantity of 108, a multiple of 12. Just-In-Time considers this excess a waste. The problem of pansize mismatch can be solved by asking the supplier of the parts to pack them in multiples of the production rates.

Safety stock. The system will allow a company to define any quantity of safety stock it wants to store on the manufacturing floor. In a Just-In-Time system, this modifier should be zero.

The Bill of Materials Records and Parts Routing

A manufacturer must store on the computer the bill of materials, specifying the repetitive components required to build the product. This allows the company not only to define the product structure but also to specify the routing of every part throughout the process. It can also specify any relative time offset with respect to other parts.

Component operation sequence number. In case the assemblies need to be routed through a sequence of work centers, the sequence numbers will tie each component to a specific work center and WIP location. This information will allow the stock location to issue a component directly to the required work center.

Component start days offset. The offset lead time specifies the number of days from the start of the assembly build cycle to the time when the part is actually needed at the work center. A manufacturer can use the offset times in a Just-In-Time system to issue the parts at the precise moment when they are needed.

Creating and Maintaining Repetitive Work Areas

MANMAN/REPETITIVE has a set of commands to define the work areas in the factory and the parts associated with them.

A work area is an area on a shop floor where workers process repetitively manufactured parts. A manufacturing process may involve a single work area or a collection of work areas.

A work area is not the same as a work center. A work area may require a routing that involves multiple operations in different work centers. In such cases, a primary code is needed to tie the operations to the work area in which they take place. The command ADD A ROUTING (AD,500) defines the primary code that defines a particular routing of work centers as a work area.

Some of the commands to handle the definition of work areas are as follows:

AD,770: Add a work area.

CH,771: Change a work area's description.

CH,772: Change a work area's default WIP location.

DE,770: Delete a work area.

LI,770: List a work area.

To use the command ADD A WORK AREA (AD,770) to add a work area to the system, there must be two pieces of information related to the command:

- the work area description, which identifies the work area
- the WIP inventory location where the components are stored during the build cycle

The next step is to establish the part–work area relationships.

MA,371: Establishes the relationship between a part and the work area that has demand for it.

RE,371: Will report all the work areas associated with a particular part or all the parts associated with a particular work area.

Then, to obtain a report of all the work areas in the system, the command WORK AREA REPORT (RE,770) is used. The report lists work areas descriptions, default

WIP locations, and clearing accounts. The report can be sorted by work area or by WIP location.

Defining and Maintaining Build Schedules

Once a manufacturer has defined the work areas and identified the repetitive parts in the process, it can add a build schedule for those parts. The build schedule defines the quantities of a part that the manufacturer plans to complete at a particular work area. Within each build schedule, the manufacturer can define the daily rate of completions, either manually or automatically.

Some of the commands to input this information to the computer are as follows:

MA,370: Maintain build schedule by part number.

MA,372: Maintain build schedule by work area.

LI,371: List a work area's scheduled production.

LI,370: List a part's build schedule.

RE,370: Build schedule report for one or all parts.

A unique build schedule exists for each work area–primary code combination for a part. A manufacturer could build more than one part in one work area or build the same part in different work areas.

In case a manufacturer wants to roll forward any incomplete build schedule, it can use the ROLL FORWARD BUILD SCHEDULES utility command (UT,369). It can roll forward the incomplete quantity to the next day's build schedule or to any future day. It also has the choice of decreasing overcompletions from future build schedules. The command will generate a report of the quantities rolled forward or deleted.

Once the build schedule is entered into the computer, the system uses the information in three different ways:

1. It communicates material requirements, capacity requirement planning, and resource planning functions to the supply sources.
2. It identifies the assemblies scheduled for production, allowing the system to calculate their starting day and which components are needed to produce them.
3. It measures scheduled output against actual output and reports the results.

Executing the Production Process

The MANMAN/REPETITIVE system controls the manufacturing floor function with four basic processes:

1. pulling components to the work centers
2. recording assembly scrap
3. monitoring production output
4. backflushing job completions with earned material and labor

Pulling Materials to the Work Centers

The MANMAN/REPETITIVE system pulls the parts for the work centers only when the workers need them. This is very different from a work order system, where the stockroom clerk kits and issues all the parts called on the work order without considering the actual time at which the work center is going to use them. In the repetitive system, a manufacturer would run the REPETITIVE PULL LIST REPORT (RE,375) on a daily basis to determine which components it has to pull to support the current build schedule. This report will determine the components that a work center needs to complete its task for the day.

The system first determines which assemblies are due for completion according to current and future build schedules. It then determines whether the assembly schedules will require components on the pull list date by using the following time window calculation:

> pull date + (assembly fixed lead time − 1) − component offset lead time + number of days' supply + additional manufacturing days

After the system has calculated the component's gross requirements, it nets this quantity against the component's quantity at hand in the WIP location that builds the assembly. The system also allows correction of variations on the schedule when actual jobs completions are different from the planned daily rate. To do this, the system considers the part's issue policy modifiers previously defined for every part. These modifiers will consider the part's pansize, issue number of days' supply, issue safety stock, and component yield.

Finally, the system generates a report listing the total quantities of components needed to support production and the WIP locations that will receive the parts.

Please note that it is very important that the workers run the transaction TR,376 to backflush the previous day's production before running the next day's pull list. This sequence of transactions will ensure that the stockroom clerks will pull the net requirements for materials.

Repetitive Example

The following example illustrates how the system arrives at a component pull list to meet the planned daily schedule. The example includes two assemblies, A1 and A2, with a common part C.

ASSEMBLY PART A1

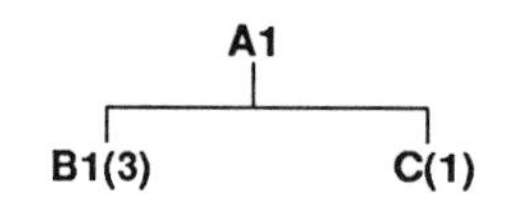

ASSEMBLY PART A2

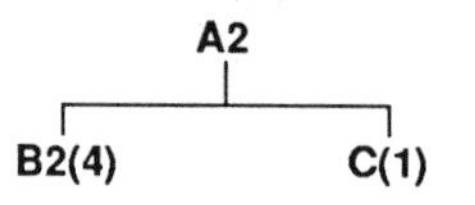

Product Structure Detail File

ASSEMBLY PART: A1

FIXED LEAD TIME: 2 days

COMPONENT	ITEM	SEQ	QPA	OFFSET
B1	1	10	3	0
C	2	20	1	1

ASSEMBLY PART: A2

FIXED LEAD TIME: 3 days

COMPONENT	ITEM	SEQ	QPA	OFFSET
B2	1	10	4	0
C	2	20	1	2

The SEQ entry identifies the work center number that has demand for the part. For example SEQ 10 means that B1s are routed to work center 10. This number is further specified in the next step.

QPA is the quantity of parts needed per assembly. For example, the process A1 will need three B1s and one C to build one assembly A1.

Finally, OFFSET is the lead time from the assembly A1's and assembly A2's start dates to the actual dates when the work center will need the parts. The Product Structure Detail File shows that it will take two days to build assembly A1. This is the fixed lead time. The work center will start building A1 using B1s and will not need part C until one day later. For assembly A2, the work center will need part C two days after it starts to build A2 with B2s. Figure 3.6 shows the offset required for Cs against the start dates for assemblies A1 and A2. The OFFSET calculation works very well in a Just-In-Time system, for parts should be issued to the manufacturing floor exactly when needed.

Routing Detail File

ASSEMBLY PART: A1

SEQ	WORK CENTER	WIP LOC
10	ASMB-1	WIP-1
20	ASMB-3	WIP-3

ASSEMBLY PART: A2

SEQ	WORK CENTER	WIP LOC
10	ASMB-2	WIP-2
20	ASMB-3	WIP-3

The Routing Detail File shows that the same work center, ASMB-3 (SEQ 20), processes all part Cs for both assemblies A1 and A2. Also, part Cs are stored at the same WIP location, WIP-3, for either assembly A1 or A2.

The next step is to enter the build plan into the Build Schedule Detail File. This information will tell the repetitive module the build rates for the assemblies A1 and A2.

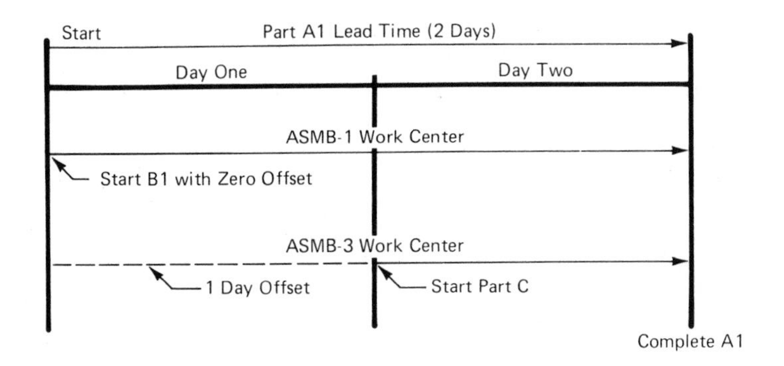

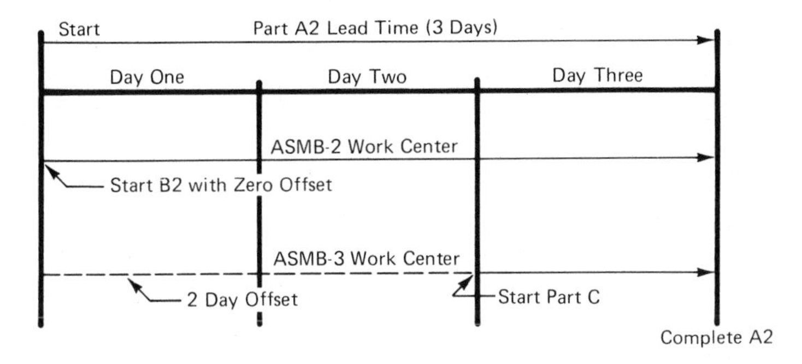

Figure 3.6 Part lead time offsets for assemblies A1 and A2.

Build Schedule Detail File

ASSEMBLY PART	MON	TUES	WED	THURS	FRI	SAT	SUN
A1	0	0	100	100	100	100	100
A2	0	0	200	200	200	200	200

With all this information on the computer, the repetitive pull list report (RE,375) will produce the pull requirements shown in Table 3.1. The completion date window is the time that it would take the part to go through the manufacturing process. For example, the stockroom issues 800 B2 parts to work center ASMB-2 (WIP location WIP-2) on Monday. These parts would go through a three-day process to complete 200 A2 assemblies by Wednesday. The QOH indicates the quantity on hand in the corresponding WIP location on any particular day.

Part Cs for assembly A2 have an offset time of two days from the starting day. It will not be necessary to pull these parts until the third day.

On Tuesday, the pulling process gets busier. The work center now needs 300 B1s,

TABLE 3.1 REPETITIVE PULL LIST REPORT

Pull list run date	Component	Completion date window	WIP location	WIP QOH	Pull requirements
MON	B1 (A1)	MON-TUES	WIP-1	0	0
	C (A1)	MON	WIP-3	0	0
	B2 (A2)	MON-WED	WIP-2	0	800
	C (A2)	MON	WIP-3	0	0
TUES	B1 (A1)	TUES-WED	WIP-1	0	300
	C (A1)	TUES	WIP-3	0	0
	B2 (A2)	TUES-THURS	WIP-2	800	800
	C (A2)	TUES	WIP-3	0	0
WED	B1 (A1)	WED-THURS	WIP-1	300	300
	C (A1)	WED	WIP-3	0	100
	B2 (A2)	WED-FRI	WIP-2	1600	800
	C (A2)	WED	WIP-3	0	200
	C (TOTAL)		WIP-3	0	300

because the lead time to build the product is two days and 100 A1 completions are scheduled to be done by Wednesday. Process A1 needs no part Cs until the next day, because of the one-day offset. More part B2s are also issued on Tuesday to build the assembly A2s by Thursday. The quantity on hand (QOH) in the WIP-2 location consists of the 800 units pulled on Monday.

Figure 3.7 shows the line-by-line phase of the part issues and completions for assemblies A1 and A2. It can be seen that after the initial start-up and pipeline filling, the system becomes repetitive with respect to issues and completions. The system will adjust any variations to the daily completion schedule, and there is always the option of selecting a day for a catch-up build schedule.

	MON	TUE	WED	THU	FRI	MON	TUE
BUILD SCHEDULE:							
ASSEMBLY A1	0	0	100	100	100	100	100
PULL SCHEDULE A1 PARTS							
B1(3)	0	300	300	300	300	300	300
C(1)	0	0	100	100	100	100	100
BUILD SCHEDULE:							
ASSEMBLY A2	0	0	200	200	200	200	200
PULL SCHEDULE A2 PARTS							
B2(4)	800	800	800	800	800	800	800
C(1)	0	0	200	200	200	200	200
TOTAL C PARTS:	0	0	300	300	300	300	300

Figure 3.7 Pipeline for A1 and A2 repetitive processes.

After backflushing all the assemblies completed on Wednesday, the balance and pull list presented in Table 3.2 results on Thursday.

After a backflush transaction, the WIP-2 location is credited with 800 B2 parts, the WIP-1 location with 300 B1 parts, and the WIP-3 location with 300 C parts. Also, a new set of parts has been issued for the next repetitive daily rate. Assuming that the workers met the repetitive build rate, they will always have at hand one day's supply of parts at WIP-1 and two days' supply at WIP-2. Notice that they process part Cs the same day, so there is no need to stock them. In a Just-In-Time system, a manufacturer tries to reduce the process time of the parts in order to reduce not only the labor but the associated WIP inventories in the production line. Chapter 5 deals with time waste, and it makes specific suggestions on how to reduce the process lead time.

Once the repetitive build schedule reaches a steady state, there are only two reasons the rate might change. First, there might be voluntary changes in the build plan. Production planners could adjust this rate change manually if the change is temporary, or the MANMAN/REPETITIVE system could change it automatically if the change is permanent. Second, the manufacturer might not meet the build plan. In the example above, the manufacturer could have built only 75 A1 assemblies and 150 A2 assemblies on a particular day. In this case, the repetitive module would again pull the same rate for the next day, netting the parts left in the WIP locations. The stockroom would not issue excess parts to the manufacturing floor. This is an excellent feature of a Just-In-Time system, but it creates the problem of catching up when there are missed schedules.

Just-In-Time provides a solution to the latter problem through the use of spare capacity. Just-In-Time would call for using the manufacturing process at less than maximum capacity. Then work centers could use the resulting spare capacity to catch up on schedules. This approach could create problems for fast-growing companies, which tend to be always behind their capacity curve. Where there is limited spare capacity, overtime is the best solution. If the schedule miss is very large, it will be necessary to rearrange the daily rate to catch up during a longer time window. It is important to understand the reason for the missed schedule and to take the necessary corrective actions to avoid a repetition of the problem. Just-In-Time gives higher priority to understanding the reason than to the actual catch-up process.

TABLE 3.2 BALANCE AND PULL LIST

Pull list run date	Component	Completion date window	WIP location	WIP QOH	Pull requirements
THURS	B1 (A1)	THURS-FRI	WIP-1	300	300
	C (A1)	THURS	WIP-3	0	100
	B2 (A2)	THURS-SAT	WIP-2	1600	800
	C (A2)	THURS	WIP-3	0	200
	C (TOTAL)		WIP-3	0	300

Backflushing Full Assemblies

The Ask MANMAN/REPETITIVE system provides two commands to deal with backflushing full assemblies:

TR,376: Backflush a part from a work area.

TR,377: Reverse the backflush of a part from a work area.

Workers must run backflush transactions at the end of each working day. The location of the backflushing transaction must be at the work center that completes the part. In the example given earlier, work center ASMB-3 completes the assemblies A1 and A2 after it assembles part C. This work center will be the logical place to do the backflushing for assemblies A1 and A2. The operator will simply count completions during the shift and execute the backflush before going home.

The backflush of full assemblies performs many functions on the system:

- The assembly earns the labor and overhead for all the operations completed at each work center. The transaction posts this earned labor to the Labor Distribution Detail File for efficiency analysis.
- The backflush subtracts all the parts that compose the assemblies completed from the quantities at the component's WIP location. The transaction also earns the standard costs of the parts and posts the inventory movement on the Transaction Log File for audit reporting.
- The backflush moves the completed assemblies to a new inventory location or to another WIP location.
- Finally, the backflush transaction allows the information just entered to be edited. This might include deleting components, substituting components, substituting WIP inventory locations, and altering quantities consumed.

The reverse backflush transaction allows the operator to correct mistakes, reversing the parts, or subassemblies, back into the previous inventory location.

Backflushing Intermediate Assemblies

MANMAN/REPETITIVE provides the capability of executing an intermediate backflush transaction for the purpose of having a count point operation. This transaction is similar to backflushing full assemblies, but it can be used on an incremental basis during the manufacturing shift. The backflushing of full assemblies, on the contrary, is only executed at the end of the shift to count total completions.

The intermediate backflush transaction can be used for incrementally counting parts completions, consumed materials, and earned labor at a particular point in the process. The system posts these values in the WIP Value Detail File, where they can be monitored at any particular time in the process. Companies that have long manufacturing cycles use the intermediate backflush to track the progress of their WIP. Also, manufacturers can use intermediate backflush to subcontract work outside the factory.

The commands to execute these transactions are as follows:

TR,370: Backflush at a count point from a work area.

TR,371: Reverse a backflush at a count point.

These count point transactions should be used as little as possible. Just-In-Time considers transactions whose only purpose is to keep track of intermediate completions to be a waste.

Repetitive and Cost Accounting

The MANMAN/REPETITIVE system uses the same basic product costing system as the MANMAN/MFG system. The system logs all the repetitive transactions onto the Transaction Log File. A report on this information is available at any time during the fiscal month and at the month's end.

The collection of the manufacturing costs is quite different from that of a work order system. The WIP Locations Accounts stores the WIP inventory values produced on the manufacturing floor. The Work Area Clearing Accounts collects the material and labor cost variances on the work area variances. When backflushes are used at count points, the system stores the intermediate costs in the WIP Value Detail File.

Material and Labor Variances

MANMAN/REPETITIVE tracks and reports material and labor variances in a way similar to a work order system. There are, however, two variances that have a lesser meaning in a repetitive system: material usage and lot size.

Material Variances

Configuration variance. This variance occurs as a result of alterations in engineering design that change the material usage. An engineering change order modifying the product so that it requires different quantities of components will create a configuration variance.

The system obtains the configuration variance by comparing the cost of the components used to build the product against the standard material and component costs stored on the computer. MANMAN/REPETITIVE uses the information stored on the material clearing account associated with the work area to calculate this variance.

Material usage variance. This variance occurs when the work center uses additional (or fewer) materials for a work order than were allocated. With MANMAN/REPETITIVE, materials are issued to a WIP location in the exact quantities required to meet the build rate. Also, materials are issued for several assemblies at the same time. Unlike a work order system, there is no concept of lots, for it is difficult to determine the difference between the scheduled quantities and the quantities actually issued for a given assembly or work area.

Purchase price variance. The system calculates this variance for all the inventory receipts vouchered during the accounting period. This variance is the same for both the MANMAN/REPETITIVE system and work order systems.

Labor Variances

To report labor variances, a manufacturer needs to record the actual labor by work center and maintain routings for the parts that it manufactures.

Rate variance. This variance occurs when the labor rates of employees differ from the standard labor rate in the work center in which they report their hours. The rate variance is the difference between the actual labor rate and the work center labor rate times the reported hours.

Efficiency variance. This variance measures the numbers of hours that an employee works more (or less) than the standard number of hours expected in a particular operation. The system can report efficiency variance by work center, which it calculates by multiplying the difference in labor hours by the standard work center's labor rate.

Lot size variance. This is the variance that tracks the difference in setup cost per unit when the planner orders a nonstandard lot size. It is associated with work order quantities and is meaningless in a repetitive system.

Methods change variances. This variance occurs when a routing for a part differs from the one used to develop its standard cost. For example, if a company adds a new operation to the process or deletes an existing one, the total labor hours for that routing would change. The repetitive system calculates this variance at the time of a final, intermediate backflush or a complete backflush.

It should be noted that all the above variances will help a manufacturing company to track any process deviation from standard values. But the tracking and collecting of all that information entails that the company must have workers and overhead staff executing the transactions and the analyses. In a Just-In-Time system, these activities must be minimized, for they don't directly add value to the product.

The best combination is to set up a simple system that only captures key measurements on the process's efficiency. This data collection system should not burden the workers or require a large overhead staff to maintain.

3.4 REPETITIVE MANUFACTURING AND PARTS SHORTAGES

In a work order system, the materials planner opens a work order on the computer and releases the information to the stockroom. The clerk at the stockroom asks the computer to print a component availability check and a kit picking list. If all the parts are in the stockroom, the clerk will pull the materials and send the work order to the manufacturing floor.

Let's assume that the stockroom clerk has a dozen work order kits to pull and many different parts shortages. The computer system will print a parts shortage report that lists all the parts required to complete those kits. Then the materials department uses this report to expedite getting those parts. Hurrying parts when they are needed is very inefficient and slows down the production process. It also indicates there are operational problems within the materials department. For example, perhaps there was a late release of a purchase order and the part's supplier was not given enough lead time to deliver the part. On the other hand, a shortage might indicate that the supplier doesn't meet delivery commitments or ships poor quality parts. In any event, the length of the shortage list indicates that something is wrong in the system.

To compensate for the delay caused by short parts, materials departments tend to release work order paperwork to the stockroom with enough lead time to fill shortages. This process creates the waste of excess inventory and additional labor overhead. Nevertheless, the work order system allows the materials planner to get ahead of the schedule to fill shortages.

In a repetitive system, there is no way to know which parts are short until the day the stockroom clerk pulls the kit to issue the parts to the floor. This is the wrong time to find out that there is a shortage. The problem is created because the repetitive software cuts the order to pull parts the same day that the work center needs them. Since the daily build schedule must net the current rate against the balance left from the shift the day before, there is no way to know in advance the specific parts requirements.

One way to solve this problem is to have a rolling demand report that includes the repetitive schedules for a short production window. This report then checks the stock and WIP locations for the parts that are at hand to meet this repetitive schedule and lists the possible shorts that might be encountered. Whatever the rolling production window, the report will list the additional parts that are going to be needed during that time. As supplemental information, the report can give the status of the open purchase orders to procure those parts.

This daily report gives the materials planners and buyers a peek into the immediate future, pointing out areas of trouble before they occur. Figure 3.8 shows a sample page of such a report for Ask MANMAN/REPETITIVE. This report is used at Metaphor Computer Systems to monitor possible shortages in the production line. The report was written using QUIZ, a reporting language that Ask makes available in their system.

Ask MANMAN/REPETITIVE offers a series of commands to report, record, and relieve a repetitive shortage. The system flags any shortages at the time the stockroom clerk creates a pull list report. The pull list details the components needed for all WIP locations for a particular date. The MANMAN/REPETITIVE shortage commands are as follows:

QRE951
03/17/88
MRP 03/16/88 M2: .750 M3: .000

METAPHOR COMPUTER SYSTEMS
ALL SHORTAGES BY P/N

PART NUMBER	DESCRIPTION	SC	QUANTITY ON DEMAND	QUANTITY SHORT	QUANTITY ON HAND	STK OUT DATE	VENDOR	P.O. NUMBER	QUANTITY REC/INSP	QUANTITY DUE	PROMISED DATE
58-1336-41	CAP,TANT,33UF,10V,10	BH	2	2	0	03/17/88			0	0	10/30/71
65-0102-01	TERMINATOR,COAX-BNC,	B	11	11	0	03/17/88	INMAC	CLOSED PO	0	0	
							INMAC	PY803032	0	26	04/04/88
65-0106-01	ADAPTER,THIN BNC-F	B	22	6	16	04/07/88			0	0	10/30/71
71-0123-01	ASM,CBL,F/S DISK POD	BH	19	6	13	04/06/88			0	0	10/30/71
71-0140-01	CABLE, M	BH	16	3	13	04/10/88	WESTERN RE	CLOSED PO	0	0	
							HAMILTON /	CANCLD PO	0	0	
71-1009-00	CABLE ASM., RECHARGE	B	110	9	101	04/13/88	DE ANZA MF	PR710080	0	100	03/07/88
							DE ANZA MF	PR710080	0	100	03/28/88
							DE ANZA MF	PR710080	0	100	05/16/88
							DE ANZA MF	PR710080	0	100	06/13/88
							DE ANZA MF	PR710080	0	100	07/18/88
							DE ANZA MF	PR710080	0	200	03/16/99
71-1013-00	CABLE ASM., VIDEO, W	BH	110	55	55	04/01/88	ICONTEC IN	CANCLD PO	0	0	
71-1018-01	CABLE ASM, XMTR RCVR	B	10	10	0	03/17/88	BIL-MAR AS	PR803004	0	5	03/21/88
							BIL-MAR AS	PR803004	0	5	05/02/88
71-1020-00	CBLE ASM, 60 POS.-DI	BH	6	6	0	03/17/88			0	0	10/30/71
71-1021-00	CBLE ASM., 26 POS.-D	BH	11	11	0	03/17/88			0	0	10/30/71
71-1027-00	CABLE ASM,W/S1,CRT/B	BH	52	52	0	03/17/88			0	0	10/30/71
71-1102-00	CABLE ASM., AC PWR -	BH	110	14	96	04/12/88	HAMILTON /	CANCLD PO	0	0	
75-0050-01	FAB, PCB W/S CPU	B	2	2	0	03/17/88	CIRCUIT VI	CLOSED PO	0	0	
75-0051-00	FAB,PCB,W/S MEMORY	B	8	8	0	03/17/88	WEST COAST	CLOSED PO	0	0	
							AMTRONICS	CLOSED PO	0	0	
							CIRCUIT VI	CLOSED PO	0	0	
75-0057-00	FAB,PCB,RECHARGE	B	18	17	1	03/18/88	SIGMA CIRC	CLOSED PO	0	0	
							SOUTH BAY	PR802001	0	100	04/11/88
75-0074-00	PCB FAB, FS DISK CTR	B	26	14	12	03/30/88	ADVANCE CI	PR704004	0	68	03/21/88
75-0110-00	FAB,PCB,WS1 24V REG	B	68	23	45	04/05/88	SOUTH BAY	PR801047	0	75	03/28/88
78-9016-00	FUSER ASSEMBLY LASER	F	3	3	0	03/17/88			0	0	10/30/71
82-0051-02	ASM., W/S I MEMORY P	MS	11	5	6	04/02/88			0	0	10/30/71
82-0057-03	ASM., RECHARGE PCB	MS	43	43	0	03/17/88	VANTRONICS	PR803007	0	25	03/17/88
82-0074-00	PCB ASM., FS DISK CT	MS	19	6	13	04/06/88	VANTRONICS	CANCLD PO	0	0	

Figure 3.8 Sample of a rolling window parts' shortage report for a repetitive system.

LI,368: List a repetitive shortage.

RE,368: Report a repetitive shortage.

TR,368: Record a repetitive shortage.

TR,369: Relieve a repetitive shortage.

3.5 MATERIALS PULLING AND REPETITIVE MANUFACTURING

A common complaint against repetitive manufacturing systems is the additional labor incurred in the daily pulling of parts to feed the production line. In a Just-In-Time system, the task of pulling parts from stock locations doesn't add any value to the product and should be minimized.

There are a few things that can be done to reduce the labor associated with pulling parts, making the system simpler and less prone to mistakes by stock clerks.

First, the repetitive parts can be classified into three groups on the basis of the frequency and method of pulling.

Group 1. This group includes small parts (e.g., screws, washers, small plastic parts, wires, and cables) that can be released to the manufacturing floor in a bulk issue. The stock clerks can issue these parts weekly, biweekly, or monthly. They account for a high volume of usage but a very low total dollar value. Normally a manufacturer would expense these parts at the time of issue to a separate inventory account to minimize financial tracking.

Group 2. In this group are parts that are pulled for a daily schedule and then dispatched to the work center at the beginning of the work shift. For high-volume operations, the stock clerks can dispatch the parts in many installments during the day. The manufacturing floor must have a location to store a full issue of these parts. For example, if the stock clerks are issuing parts every two hours, there must be a two-hour storage allocation available at the work center that receives the parts. For a typical product, these parts should constitute about fifty percent of the parts required to build the product. A Kanban system is ideal to coordinate the issuing of these parts so as to avoid overissues when the work center is not consuming them at the scheduled rate. Chapter 4 reviews the Kanban principles in detail.

Group 3. This group includes bulky parts that the stock clerks will issue daily but, because of the large space they take, only when there is a demand request. These parts are dispatched one by one, and a Kanban system is ideal for coordinating their release. One simple method is to use an intercom system; the line worker calls the appropriate stock location to order a new one. However, this creates the problem of time waste at the work center, for the worker must have a buffer of one part at hand to eliminate the idle time between deliveries.

Stock Location Routing and Parts Packaging

There are some other methods for reducing the time required for pulling parts from stock locations. These methods vary depending on the type of part, its size, and its compatibility with different processes and work areas.

One simple but effective way to improve the part counting is to ask suppliers to pack their parts in multiples that make the counting simple. For example, if stock clerks are issuing a part in quantities of 500, the quantities will be easier to count if the supplier packs the part in exact lots of 250. Making such arrangements with suppliers normally takes time, and the production engineers should study the options carefully to obtain maximum results.

Improvements are also usually possible in the physical arrangement of the parts in the stock locations. The objective is to minimize the stock clerks' traveling distances during the process of pulling parts. This requires ordering the parts in the stock location by their relative sequence on the pull list rather than by their part number order.

In general, the idea is to encourage the materials staff to optimize the pulling time. After all, Just-In-Time considers the task a waste.

The Stockroom and Stock Locations

As noted, there is a distinction between stockrooms and stock locations. Stockrooms, in a typical manufacturing environment, are secured, usually fenced, locations where materials are stored for a long period of time. Stock locations, on the contrary, are staging areas, usually open, where materials are stored temporarily until they are transferred to the manufacturing process. Stock locations should contain only a few days, or hours, of supply. Stockrooms might contain materials that are there for weeks or months.

In a Just-In-Time system, stockrooms are a waste and should be of minimal size. Just-In-Time looks at stock locations as small, dynamic buffer locations for materials on their way to the production process. For example, a company could buy large quantities of small parts, usually type C parts, that are issued to work centers in bulk issues and it could store them in a stockroom. Then, it would be able move parts that it issues daily directly from the receiving dock to a stock location in front of the work center materials dispatch area. This stock location might be just a square painted on the floor.

3.6 REPETITIVE MANUFACTURING AND MANUFACTURING PULL SYSTEMS

There is very little conflict between the concepts of a repetitive manufacturing system and a pull system. A repetitive manufacturing system conforms to the rules of a pull system as long as the system issues materials to the production line on a netting basis.

MANMAN/REPETITIVE looks at the production completions at the end of the previous shift before issuing new parts to meet the production rate of the next shift. This procedure compensates for cases where the workers output fewer completions during the shift.

Even in the worst case (zero completions), the software will check the WIP inventory on the floor and will pull no new parts until the process consumes them. This is in essence a pull system in which parts are issued based on their rate of consumption.

3.7 SUMMARY

Repetitive manufacturing uses a simpler method to issue and track materials on the manufacturing line than the one used in a work order system. But that is not the only advantage of a repetitive system. The system also forces a manufacturer to have linear productions rates and causes less material waste than a work order system.

Quality is another factor that bears on whether to use a repetitive system. A manufacturing company with a linear production rate is capable of producing a better quality product than one that has to rush at the last hour to get products out. The pressure on the workers and supervisors is higher when there is a rush, and they are more prone to make mistakes.

The implementation of a repetitive manufacturing system is not a simple task. A manufacturer must do some homework before it starts making changes in its production lines. It must design the system well and install the right software to handle the operational aspect of the system. Also, it must have a good training program to make sure that the workers and supervisors understand how to operate in the new environment. It must clearly define the stock-pulling procedures and classifications determining which parts to pull under which rules (e.g., which parts are bulk issue, daily pull, or pull on demand). In any event, the rules for material pulling described in section 3.1 apply to the repetitive environment regardless of the software system acquired to run the factory.

The final trap to avoid in using a repetitive manufacturing system is that a high volume is required. The truth is that such a system will work well at any volume, including one assembly per shift. Repetitive principles are not volume dependent. They are procedural-oriented. *Remember, the ideal production quantity in a Just-In-Time system is one.*

REFERENCES

ASK COMPUTER SYSTEMS. *MANMAN/REPETITIVE Manual,* REL 6.0, 1987.

CONSTANZA, JOHN R., and DAVID R. WAGNER. "Rate Based Planning and Material Requirements Planning." In *Just-In-Time Manufacturing Excellence.* Published by authors, 1986.

HALL, ROBERT W. *Zero Inventories.* Homewood, Ill.: Dow Jones-Irwin, 1983.

HERNANDEZ, ARNALDO, and EDWARD LOIZEAUX. "Manual to Inhouse: A Smooth System Transition." *INTERACT Magazine,* November 1986, pp. 56–61.

KIRK, STEPHEN A. "Just-In-Time in a Repetitive Manufacturing Environment." In *MANMAN Conference Proceedings, March 1987.* Los Altos, CA: pp. 222–239.

MCGUIRE, KENNETH J. "Japanese Production Planning: A Successful Balancing Act." *CIM Review: The Journal of Computer-Integrated Manufacturing Management* 1 (Spring 1985): pp. 59–64.

4 ★

Pulling Materials in the Factory: The Kanban Concept

In Chapter 2 there was an explanation of the difference between a push and a pull manufacturing system. This difference was illustrated using the example of a row of dominoes representing a manufacturing process and the movement of materials through it. In a push system, we roll down all the pieces—move the materials and build the product—by pushing the first piece in the row. In a real process, this is equivalent to a company continuing to push materials throughout the process. This activity normally continues even if the company is not consuming the materials at the same rate that it is issuing them. Once started, this process is very difficult to stop because of the dynamics of the system. The people using it would not react quickly to sudden changes in the demand for a part.

In a pull system, all the dominoes are tied with an invisible string. We begin by pulling down the first piece, but when this piece goes down, the ones behind it remain standing. We then have the choice of pulling the next one using the string. In this method, we can stop the pulling process anytime, and the remaining dominoes will stay up without any trouble.

Let's see how to implement this concept on an actual production line. Suppose that a manufacturer gives the production schedule to the work center that completes the last operation in the production process. This is the work center that sends the completed parts to the finished goods area. Imagine that the supervisor in the work center learns that she has to build one hundred units of a particular product that day. She needs one hundred sets of whatever parts and subassemblies are used to build the product. Then the supervisor checks the parts that she has at hand. If she doesn't have enough of them, she will send a request to the previous work center that supplies the parts asking for the exact quantities needed to build the product. This process creates a reverse rippling effect throughout all the factory.

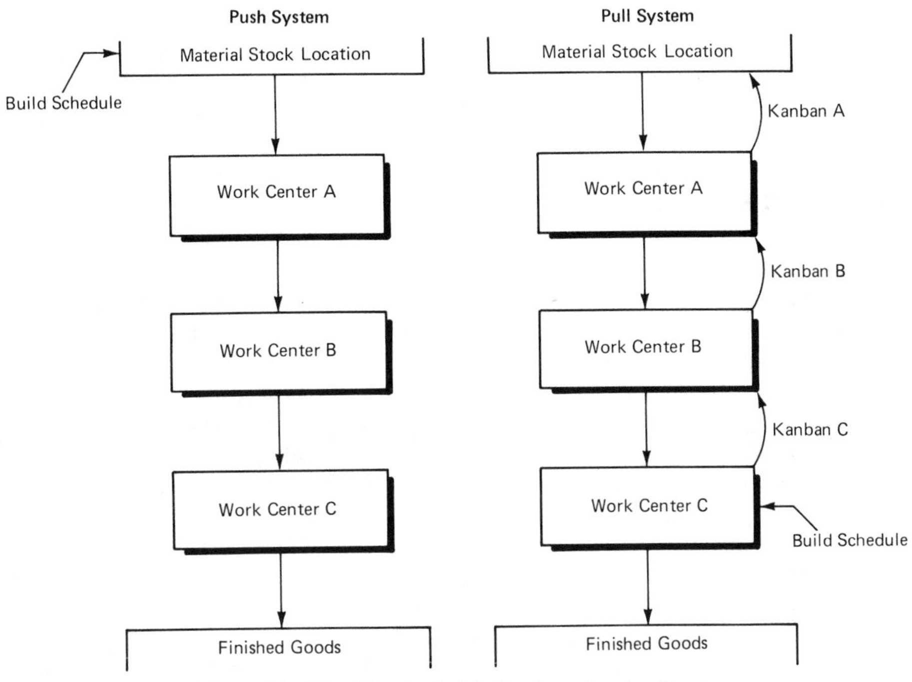

Figure 4.1 Material and schedule flow in push and pull systems.

The system always pulls the parts and subassemblies through the process towards the last work center.

This type of system modulates itself in response to variations in the production rate during the day, avoiding excess of materials in work centers. Also, if a serious problem arises which demands a line stop, the system will react quickly.

In this system, a work center requests materials from another using a card called a Kanban, and the system is called a Kanban system. A Kanban is equivalent to the invisible string that pulls down the dominoes.

Figure 4.1 shows the difference between a push system and a pull system. The push system is a single-flow process in which build schedules and materials travel in the process in the same direction. In the pull system, there is a double flow. Materials travel in one direction and the build schedules travel in the opposite direction. A Kanban system is used to communicate this schedule from one work center to another.

4.1 KANBAN AND THE TOYOTA SYSTEM

Toyota used the Kanban system widely for many years as a means to communicate materials needs between two process centers. *Kanban* in Japanese means *visible record.* Facing the challenge to eliminate waste, Toyota developed the Kanban system to move materials in a controlled environment driven only by the usage of parts.

The Toyota production system was the result of an evolutionary process that lasted many years. It was not until the early seventies that Toyota gained proper recognition for reducing its inventory down to levels not even dreamed of at the time. Toyota credits a large portion of its success to the use of a Kanban system.

4.2 SUBSEQUENT AND PRECEDING PROCESSES

Before proceeding to a review of the Kanban system, it is important to understand what subsequent and preceding processes are. These processes are used to define the rules that govern the movement of Kanbans. A subsequent process for a particular process might be a preceding process for another one, depending on their relative positions in the manufacturing flow.

To visualize the concept of manufacturing flow, imagine a series of streams moving slowly toward the ocean. These streams sometimes join others to form more robust ones, but they always keep moving. The material flow is analogous to the water running down the streams and heading toward the ocean, which represents the final destination—the customers.

Pick any place along a stream, there is always another place downstream that receives the water passing through the place picked. There are also places upstream sending water down to the place picked. Finally, each stream has a source and an end where it runs into the ocean. In a Kanban system, the sources are equivalent to material suppliers. The ends of the streams correspond to the last operations in the manufacturing process before the products reach the finished goods location.

Subsequent Processes

Assume now that we are standing in front of a work center that processes parts for a particular product. In normal production, after the work center finishes the parts, the operator will send them to another work center for the next step. The manufacturing process downstream, where the current process sends its parts, is called the subsequent process. The work center that receives the assembled parts is the subsequent process to the process that assembles the parts.

Let's assume now that the parts that arrive at the secondary assembly process are coming from another work area, which produces them using injection-molding machines. The process that is assembling parts will be the subsequent process to the upstream injection-molding process.

Preceding Processes

To continue the example, let's suppose that we walk to the process that receives the assembled parts and look upstream at the process that is assembling them. This process will be the preceding process to the process that we are facing now.

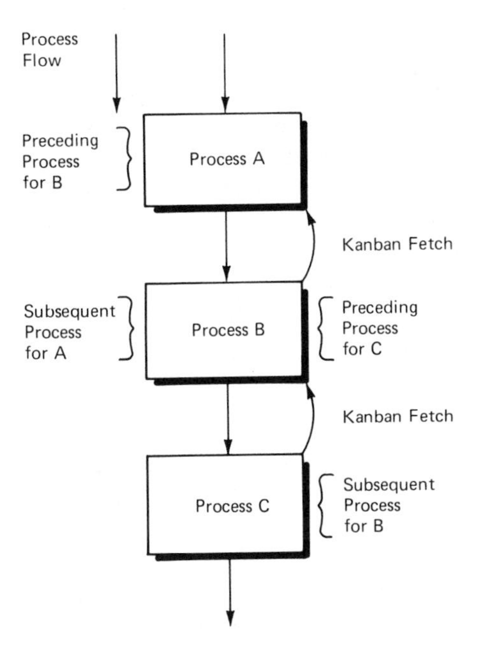

Figure 4.2 Subsequent and preceding process flow.

A Kanban will always fetch parts from preceding processes and it will always send parts to subsequent processes. Figure 4.2 shows this occurring with respect to three processes, A, B, and C. Process A sends its parts to process B and is therefore the preceding process of process B. Conversely, Process B is the subsequent process of process A. Since process B sends its parts to process C, process B will be the preceding process of C and C will be the subsequent process of B. Note that there are not exclusive relationships between processes: A process can be a subsequent or preceding process of many other processes at the same time.

To sum up, a Kanban system consists of a set of cards that travel between preceding and subsequent processes, communicating what parts are needed in the subsequent processes.

4.3 WITHDRAWAL AND PRODUCTION KANBANS

In general we have talked about Kanbans without distinguishing types. A Kanban system will require two types of Kanbans to operate correctly: a withdrawal Kanban and a production Kanban. This section focuses on both types. Note that the rules given in Section 4.4 apply to each type of Kanban independently of function. Also, the two types of Kanbans do not differ in physical appearance, but a label indicating the type should appear in large print at the top of each card. One way to differentiate Kanbans by type is to use different colors; thus workers will know immediately which is which and be able to avoid the mistake of mixing them.

Withdrawal Kanbans

The withdrawal Kanban travels between work centers and is used to authorize the movement of parts from one work center to another (Figure 4.3). In a Kanban system, a withdrawal Kanban must always accompany the flow of material from one process to another.

A withdrawal Kanban must specify the part number and revision level. It also must specify the lot size and the routing process. The Kanban must show the preceding process name and its location in the building and the subsequent process and its location.

Once a withdrawal Kanban fetches parts, it will stay with them all the time. Then, after the subsequent process consumes the last part of the lot, the Kanban will travel again to the preceding process to fetch new parts.

Production Kanbans

As mentioned before, a withdrawal Kanban communicates between work centers, passing authorizations to move parts between them. The production Kanban's job is to release an order to the preceding process to build more parts.

When the withdrawal Kanban arrives at a preceding process, most likely it will find available one or several containers with the parts that are to be fetched. A production Kanban must escort the containers at the time. The employee servicing the work center will attach the withdrawal Kanban to a visible place on the containers and then send them to the subsequent process. Before moving the containers, the employee will retrieve the production Kanban. This Kanban authorizes the work center to build a new lot of parts.

The production Kanban will go into a queue with other production Kanbans at the work center. After the work center builds the new parts, the production Kanban will travel

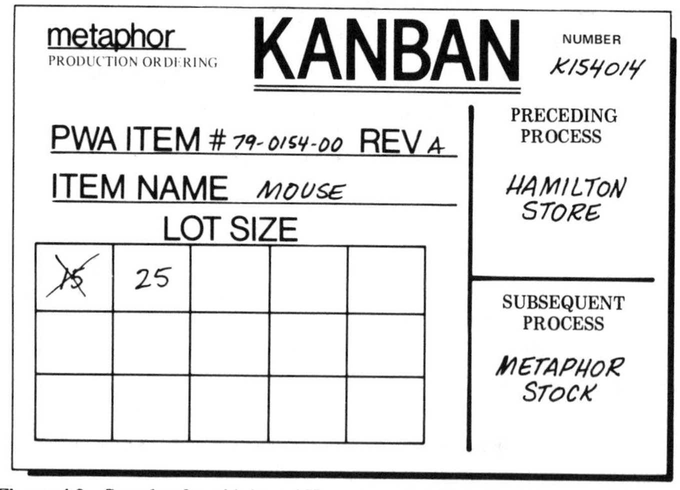
metaphor
PRODUCTION ORDERING
KANBAN
NUMBER
K154014
PWA ITEM # 79-0154-00 REV A
ITEM NAME MOUSE
LOT SIZE
~~15~~ 25
PRECEDING PROCESS
HAMILTON STORE
SUBSEQUENT PROCESS
METAPHOR STOCK

Figure 4.3 Sample of a withdrawal Kanban used at Metaphor Computer Systems.

back to the wait area until a new withdrawal Kanban starts the cycle over again.

Figure 4.4 shows the interaction between production and withdrawal Kanbans at three work centers. Notice that there is a staging area where the production Kanbans wait for the withdrawal Kanbans to arrive to fetch parts. Also notice that withdrawal Kanbans feed the parts into the work center's process exactly where they belong. Conversely, as the work center consumes parts to build its products, its production Kanban will generate withdrawal Kanbans at the work center's preceding process.

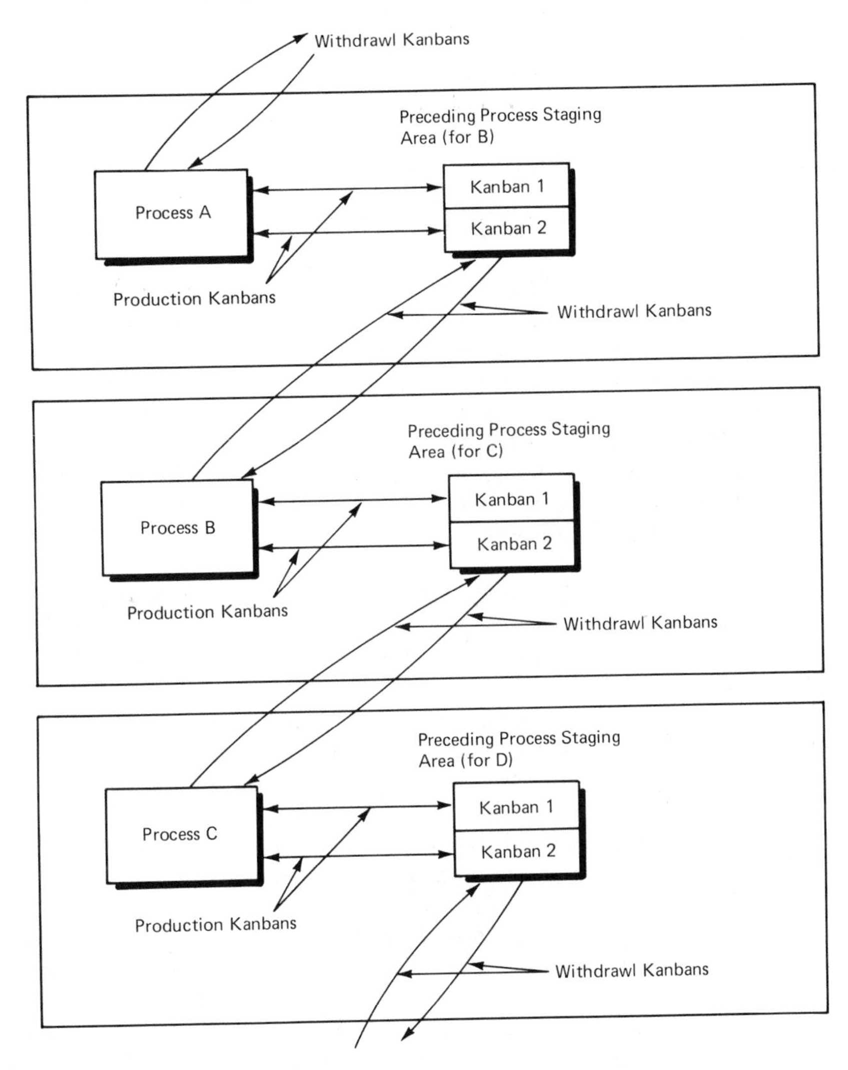

Figure 4.4 Coordination of production and withdrawal Kanbans in three work centers.

Production Kanban Shortages

In a real operating environment, there is a possibility that when a withdrawal Kanban arrives at the preceding process, there is no production Kanban waiting with the parts. In this case, the system should treat the situation as a parts emergency. The employee should send the withdrawal Kanban directly to the production area and treat it as a temporary production Kanban. The presence of a withdrawal Kanban in the work center area will give it higher priority than regular production Kanbans (but not higher than other withdrawal Kanbans there already).

4.4 KANBAN BASIC RULES

This section covers the seven rules that control the operational environment of a Kanban system. These rules are simple but very important. Any violation of them will cause disruptions in the system, with corresponding waste in materials and labor.

The first step in implementing a Kanban system is to put in writing the operational rules that control the system. It is recommended that workers involved in the system clearly understand these rules before starting to use the Kanbans. It is also important that once the Kanban system is in place, there is a periodic review to analyze whether the workers are deviating from them. If deviations occur, the reasons for the deviation should be uncovered and corrective action should be taken immediately. The Kanban rules are in tune with Just-In-Time system principles. They provide a major contribution to the avoidance of excess inventory on the manufacturing floor.

Rule 1: Move a Kanban only when the lot it represents is consumed. This rule calls for the subsequent process to withdraw the necessary parts from the preceding process in the necessary quantities and at the exact time needed. A subsequent process must send a Kanban to the preceding process asking for more parts only after the subsequent process has consumed all the parts that the Kanban was escorting.

For example, suppose a Kanban card escorts a container box with fifty parts. During a shift, the process consumes those parts one by one. That process sends the Kanban to the preceding process when it has used the last part in the box. The problem of idle time at the work center is created if there are no more parts available. How to do scheduling under a Kanban system will be discussed in a later section of this chapter.

Rule 2: No withdrawal of parts without a Kanban is allowed. A Kanban is the only vehicle for allowing the movement of parts from a preceding process to a subsequent process. No preceding process can move parts without the authorization of a Kanban request.

This rule requires a lot of discipline on the part of line workers, because it is very easy to break. They must be made to understand that the system allows no movement of materials without a Kanban card.

Rule 3: The number of parts issued to the subsequent process must be the exact number specified by the Kanban. This rule is simple to understand. The preceding process must not issue Kanbans with a partial count of parts. They must have the exact quantity specified by the Kanban.

Conversely, if there is only a partial number of parts available in the preceding process, a line shortage is created. The Kanban must wait until the preceding process produces enough parts to meet the number requested by the Kanban.

Rule 4: A Kanban should always be attached to the physical products. A Kanban is a traveling card, and it will always travel alone from the subsequent process to the preceding process to request new parts. But once the worker attaches the Kanban to a new lot of parts, the card must travel with the lot until the last part is used. Then the Kanban can return again to the preceding process to fetch a new lot.

The Kanban should be attached to the lot in such a way that it is always visible to workers. This will simplify the task of identifying the lot part number and the quantity of parts in the container.

When sending a Kanban to the preceding process, it is usual to drop it into a mailbox. In a normal operating mode, the worker in charge of the preceding process checks that process's mailbox regularly for new demands for parts.

Rule 5: The preceding process should always produce its parts in the quantities withdrawn by the subsequent process. This is a rule of the Just-In-Time system. Processes should never overproduce parts in any quantity, for this is a waste of labor and materials.

As indicated in section 4.3, a withdrawal Kanban triggers a production Kanban at the preceding process to build new parts. The preceding process uses the production Kanbans to pace its output at the consumption rate of the withdrawal Kanbans that it is servicing. All of this activity should match the production master schedule plan.

Rule 6: Defective parts should never be conveyed to the subsequent process. This rule concerns the quality of the parts moved by the Kanban. In a Just-In-Time system, there is an absolute need to maintain a high level of quality in producing the parts and subassemblies. Remember, there are no buffer inventories to cover for defective parts. Workers must be aware of the critical need to produce and use quality parts at every step in the process. A total quality control (TQC) program is a prerequisite of a Just-In-Time system. The TQC program will monitor quality and help to ensure it is the highest possible. TQC programs will be discussed in Chapter 8.

Rule 6 requires that there be an effective way of reporting quality problems to the preceding process in order to carry out necessary corrections expeditiously.

Rule 7: Process the Kanbans in every work center strictly in the order in which they arrive at the work center. This rule is simple. When a work center has in its input mailbox several Kanbans from different processes, the operators in that work center must serve the Kanbans in the order in which they arrived. Any failure to do that will cause a gap in the production rate of one or more of the subsequent processes.

When a preceding process services many subsequent processes and the preceding process has no interruptions, the Kanban consumption rate in the subsequent processes sets the parts demand for the preceding process.

It is recommended that these seven rules be put in writing and distributed to all the parties involved in the Kanban system. Make sure that everybody understands and uses the rules during normal operation.

4.5 KANBAN MATERIALS PLANNING

In a manufacturing company, the materials planner is the person responsible for issuing the Kanban cards. The planner also determines the lot sizes the Kanban is going to fetch. Normally, the materials planner will sometimes issue additional Kanban cards to increase the production rate for a particular part. Conversely, the planner will take Kanban cards out of circulation to reduce the production schedule.

The planner, however, cannot determine the lot sizes without consulting the factory capacity and being familiar with the containers used to pack and carry the parts. For example, if the boxes to move the parts are in multiples of ten, the planner might select a lot size of thirty rather than twenty-five. The planner could compensate by issuing a different number of Kanbans.

The number of Kanbans issued for a particular part item is calculated by the following formula:

$$\text{Number of Kanbans} = \frac{\text{units daily demand} \times \text{order cycle time} \times \text{safety factor}}{\text{lot size}}$$

The *units daily demand* is the daily production rate for the part. The *order cycle time* is the time it takes to process the part or to procure a purchased item. The *safety factor* is usually a percentage increase in the number of Kanbans instituted as a precautionary measure for buffer inventories. The *lot size* is the number of parts that the Kanban authorizes to be fetched if it is a withdrawal Kanban or to be manufactured if it is a production Kanban.

Example of a Kanban Calculation

Let's assume that our production requirement for a particular part is 55 units a month. The cycle time of the part is 22 days and the lot size we want to use is 15 units.

$$\text{unit daily demand} = \frac{55}{20} = 2.75 \text{ units/day}$$

Let's assume that the process is not stable yet and that initially it would be wise to have a safety factor of 1.5. This means we are going to process 50 percent more Kanbans than actually needed.

Applying the formula, we get

$$\text{Number of Kanbans} = \frac{2.75 \times 22 \times 1.5}{15} = 6$$

This means that we need 4 Kanbans to run the process and will have 2 extra Kanbans as a buffer until the process is steady and predictable.

If the safety factor is reduced to 1.25, we will require a total of five Kanbans (four for production and one for buffer). If the safety factor is 1, the ideal Just-In-Time factor, we will require only four Kanbans.

4.6 KANBAN SYSTEM EXAMPLE

To further illustrate the Kanban concept, this section presents a detailed example of a preceding process (work center A) supplying parts to three different subsequent processes (work centers B, C, and D). Figure 4.5 shows the physical arrangement of all four work centers. The parts that the three subsequent work centers receive from work center A are not necessarily the same. Let's assume that work center A has reduced the setup time of its machine tools so that it can execute different production runs efficiently. When a production Kanban from any one of the three subsequent work centers arrives at work center A's queue, the workers can service the order with a minimum of wasted time.

The production planner is the person responsible for determining the number of withdrawal Kanbans required to meet the process cycle times of work centers B, C, and D. The planner also calculates the number of production Kanbans required for work center A. The planner must make sure that there are enough parts to service the withdrawal Kanbans coming from work centers B, C, and D.

Table 4.1 shows the number of withdrawal Kanbans required to service the three subsequent work centers. The table includes the daily demand of parts, the order lead times in days, the safety factors for the processes, the capacity of containers, and the number of withdrawal Kanbans required.

The formula shown in Section 4.4 was used to calculate the number of Kanbans required to service the subsequent work centers. Notice that the entry under the column Order Lead Time is the time that it takes the Kanbans to be serviced. In this case, it is actually the time that it takes the materials to travel from the preceding work center A to the relevant subsequent work center.

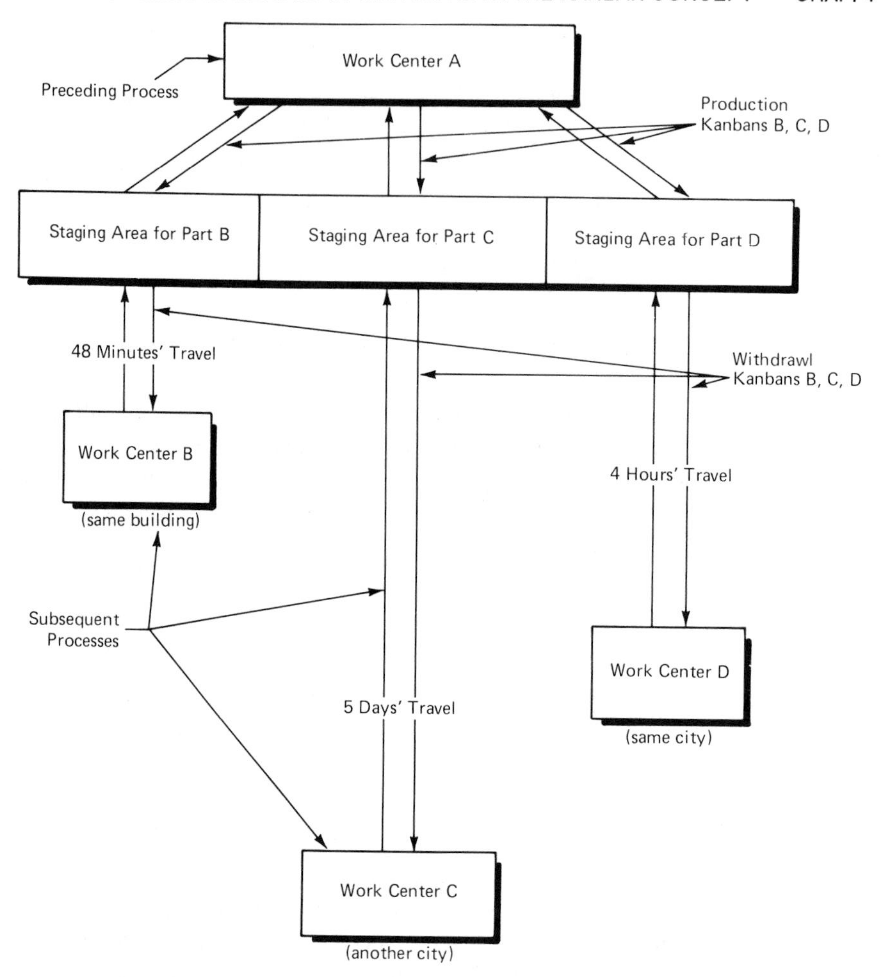

Figure 4.5 Preceding and subsequent processes arrangement.

TABLE 4.1 WITHDRAWAL KANBANS AT SUBSEQUENT WORK CENTERS

Work Center	Part	Units daily demand	Order lead time (days)	Safety factor	Container capacity	Number of withdrawal Kanbans
B	B	500	.1	1.2	30	2
C	C	100	5	1.2	100	6
D	D	60	.5	1.0	15	2

From the table opposite, it would appear that work center B is located in the same building as work center A. It takes only a little less than one hour to withdraw the parts with the Kanban.

Using the same line of reasoning, it can be assumed that work center D is in the same city as work center A. It takes about four hours to get the parts with the withdrawal Kanban. It can also be assumed that work center C is in another part of the country, for it takes five days to deliver the parts. It should be noted there needn't be a five-day return trip for the withdrawal Kanban. A simple phone call arrangement by the planner would be sufficient to release another Kanban from work center A. The withdrawal Kanban can wait for the truck to arrive at work center C's receiving dock. The receiving clerk can attach the Kanban to the materials there.

There are a few important points that need to be made. First, in spite of having a lower daily rate, part Cs (the parts used at work center C) require a higher number of Kanbans than part Bs (the parts used at work center B). The reason for this difference is the longer traveling time of part Cs. The system requires a pipeline filled with parts to ensure the uniform supply of materials to work center C. The additional parts are doing nothing but traveling, and this is a waste. The wastefulness is one of the reasons why Just-In-Time calls for selecting suppliers as close as possible to subsequent processes. The best arrangement for preceding and subsequent processes is to have the supplier located in the same building as the customer. The in-plant stores will be discussed in Chapter 6.

A nearby supplier makes practical the delivering of materials several times during a working shift, allowing smaller-sized Kanbans. For example, work center B consumes 500 parts per day, but it is only serviced with two withdrawal Kanbans of 30 parts each. Conversely, work center C, with a rate of 100 parts per day, requires six withdrawal Kanbans of 100 units each because the supply source is located five days away.

Let's assume now that the traveling time for work center B is five days and that we increase the lot size of the withdrawal Kanban to 250 units. Work center B would have to take two deliveries per day to meet the scheduled production rate. Using the formula for calculating the number of Kanbans, a total of 12 Kanbans of 250 units each would be required to do the same job as the two Kanbans of 30 units that were needed when work centers A and B were close.

Assuming that two Kanbans are at work center B in a normal working shift, ten Kanbans will be traveling during any particular day. Thus, a total of 2500 parts will be doing nothing but traveling. Under the first assumption (in which work centers A and B are close), only one Kanban of 30 parts will be traveling, and only for forty-eight minutes. This clearly shows that the best arrangement is to select suppliers as close as possible to a process—a Just-In-Time concept.

There is another factor that affects the numbers of Kanbans and their synchronization with subsequent processes. For example, a safety factor of one has been selected for part D. This factor entails that a withdrawal Kanban must be delivered on time every time parts are needed. It also entails that all the parts must be without defects.

Figure 4.6 shows the synchronization of Kanbans D1 and D2 (for work center D) using a buffer factor of one. Let's assume that at the start both Kanbans are at hand and have fif-

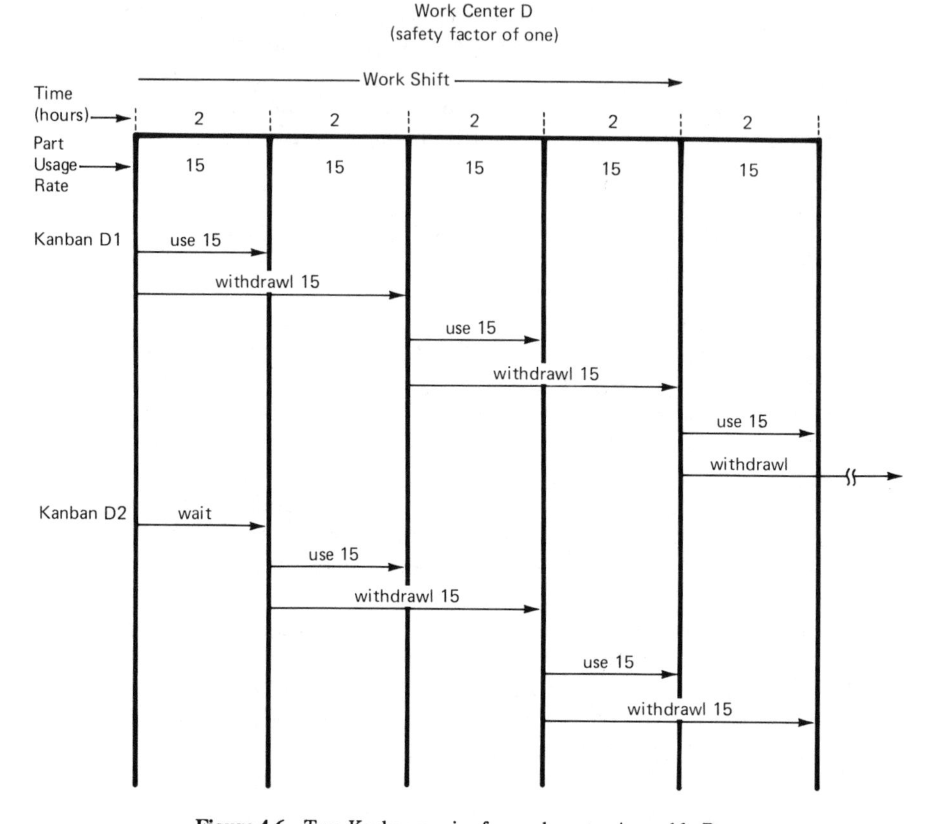

Figure 4.6 Two-Kanban service for work center Assembly D.

teen parts each. First, the fifteen parts in Kanban D1 are issued to the manufacturing floor. Then D1 is sent to work center A to withdraw another lot of fifteen parts. Note that work center D will take two hours to consume these parts. The parts in Kanban D2 will stay idle during that time. This situation, under a Just-In-Time system, is considered a waste.

After two hours, there is a need for another set of fifteen parts (assuming that the production rate is linear). Then the Kanban D2 parts are issued to the process and the Kanban is sent back to work center A to fetch new parts. Two hours later, the process consumes these fifteen parts and will need another fifteen. By that time, Kanban D1 should have been back with its fifteen parts. Kanban D1 is also sent back to work center A to fetch another set of fifteen parts. Figure 4.6 shows that the process continues, with the parts arriving at the time they are needed.

One of the obvious risks in having a safety factor of 1 is the need for perfect performance. Not only must the parts be received on time but they must also be completely unflawed. The system has no room for problems or delays.

In a Just-In-Time system, the elimination of buffers is the ideal. The claim is that, as a result, problems in the process and the system will become very clear to everybody because of the disruption they cause. Also, the system forces workers to fix problems, since only then can the process be continued.

There is no point in jumping into the swamp to drain it without being prepared to fight the alligators. Quality, process, and delivery problems don't get fixed overnight. A company implementing Just-In-Time must develop corrective plans and must work hard to execute them whenever necessary.

One simple way to avoid a process stop resulting from lack of parts is to use, temporarily, a buffer Kanban as a safety factor until the system runs smoothly. The materials planner could use a safety factor of 1.5 in the formula, which will call for three withdrawal Kanbans for work center D (two Kanbans for the required parts and one as a buffer). The parts in the third Kanban will be idle most of the time, waiting for just-in-case deliveries. This, of course, is a waste.

There are other ways to minimize the waste of buffer parts. For example, the planner could reduce the lot size of withdrawal Kanbans to 10 units so the buffer Kanban will be more economical. As Figure 4.7 shows, it will take three trips per shift to work center A to

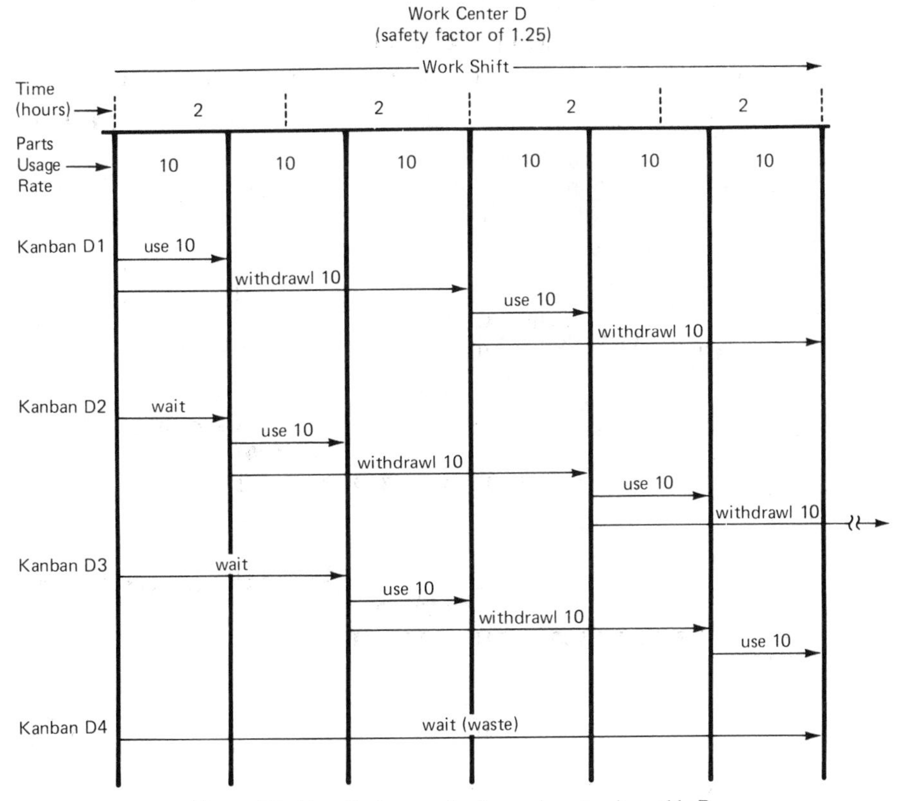

Figure 4.7 Four-Kanban service for work center Assembly D.

meet the parts requirement. The fourth Kanban would be idle at all times, with a waste of ten parts always unused. The penalty is one extra daily trip to the preceding work center. Thus, 5 idle parts having been traded for one trip. Just-In-Time considers both overhead in trips and overhead in parts a waste, and they should be minimized.

It should be clear that a materials planner has many elements to consider before selecting safety factors and Kanban lot sizes. The planner must take into consideration the quality track record of the preceding process, transportation, the linearity of the subsequent process output, and consistency in deliveries. All this information will come from different departments in the planner's organization and should be readily available during decision making.

To summarize, parts transportation between preceding and subsequent processes is of vital importance for eliminating waste in a Just-In-Time system. This subject will be covered in more detail in Chapters 6 and 7.

Preceding Process Example

We will continue now with our example, taking a look at the preceding process—work center A. This work center has three subprocesses producing parts for work centers B, C, and D. Table 4.2 shows the different manufacturing cycle times and safety factors for each work center. The table also shows the number of production Kanbans required to support the demand rate of the parts.

The lead time of each part on the table is the time it takes to manufacture the parts in the corresponding subprocess. Also, the daily production rates for each part reflect the requests by the corresponding work center. The subprocesses then determine individually safety factors, container capacities, and numbers of Kanbans.

For example, the subprocess for part B requires 9 production Kanbans to meet the production rate of 500 parts. In contrast, only two withdrawal Kanbans are used to bring those parts to work center B. Furthermore, it takes two production Kanbans to meet the demands of the six withdrawal Kanbans for work center C. Finally, it takes six production Kanbans to serve the two withdrawal Kanbans for work center D. There is not a one-to-one relation between production and withdrawal Kanbans. The only limiting factor is that both types must meet the production rates dictated by subsequent processes.

Note that it is helpful to have one type of Kanban be an exact multiple of the other. This will avoid having partial Kanbans waiting at the staging areas. For example, one

TABLE 4.2 PRODUCTION KANBANS AT WORK CENTER A

Work Center	Part	Units daily demand	Order lead time (days)	Safety factor	Container capacity	Number of production Kanbans
B	B	500	5	1.1	300	9
C	C	100	1	1.0	50	2
D	D	60	3	1.0	30	6

production Kanban with a quantity of 300 will serve 10 withdrawal Kanbans of 30 units each. If a quantity of 250 had been selected for the production Kanban, there would usually be an odd number of parts left in the container waiting for the next production Kanban to arrive.

4.7 KANBAN AND MRP II

Material Requirement Planning (MRP) is a software system designed to calculate the materials requirements for a particular time horizon. The system uses as input a production master schedule and released bills of materials to explode the materials requirements called for on the master schedule. The MRP then nets the component requirements with the inventories at hand and the open orders placed with suppliers. The system outputs the additional purchases necessary to meet the parts needs and plans the work orders required to meet the production schedules.

One of the initial difficulties with MRP was that it didn't take into consideration the plant capacity, creating problems with the scheduling of the manufacturing resources. This problem was corrected by introducing a feedback control loop to the software so that the system could compare the manufacturing workload to the capacity of the process. This new updated software was called a closed loop MRP.

Manufacturing Resource Planning (MRP II) is a software system designed to manage all the resources of a manufacturing organization. MRP II is a process that links manufacturing with the finance and sales departments, providing tools for joint decision making among all three departments. The system starts with a sales forecast and translates it into a production master schedule, which drives the material planning, the factory capacity requirements, and the production process with capacity and priorities. MRP II also provides simulation capabilities to the system to allow the evaluation of different financial alternatives.

One of the first questions facing some companies planning to implement a Kanban system is how the Kanban system would work with their MRP II systems. In some cases, a company will even raise the question whether the Kanban system could replace its MRP II system.

MRP II is a process that gives manufacturing, sales, and finance organizations a global picture of materials, capacity, and finance needs to meet the company's sales forecasts. Kanban, on the contrary, is a bottom-up process that has a very limited ability to generate an overall picture. The only information the Kanban offers to the materials planner is material needs between processes based on actual consumption. A Kanban system is like a long chain in which the individual Kanbans are the links. There is no way that a particular link can provide information about the length of the chain.

In a Kanban system, the MRP II system can provide top-down planning and financial visibility to the different processes in the system. MRP II output can also be used to forecast the monthly build for the factory on a process-by-process basis. This information can be given to the workers in charge of the work centers. But these workers must use this information only as a production forecast. The actual output commitments should be based on the demands of the withdrawal Kanbans.

Such a system needs a starter work center that knows the actual daily production schedule. This problem is solved by giving the last process the actual completions it must finish every day. This process will immediately start building the product and will use the withdrawal Kanbans to communicate its material needs to the preceding processes.

Toyota uses a similar arrangement. It gives the car completions required for the day to the last department in the line. This department then uses Kanbans to pull the material along the chain. Toyota also gives monthly goals to the other departments. These goals are used only as a forecast for capacity planning.

4.8 KANBAN AND REPETITIVE MANUFACTURING

In a manufacturing organization, there are two inventory systems running in parallel. One is the actual physical inventory of parts, subassemblies, and finished products that are stored in stockrooms and along the manufacturing process. The other is the perpetual inventory records stored on the computer data base that tracks the manufacturing process.

The computer inventory is a data inventory that should match the physical inventory up to the last part. The MRP II uses this computer inventory to calculate the component requirements to support production's material needs. The finance department also uses the computer inventory to keep track of the dollar value of the physical inventory. A disciplined manufacturing company maintains a one-to-one correlation between the inventories, transacting every movement, usage, scrap, purchase, and shipment of finished products throughout the organization.

A Kanban system moves materials in a pull fashion, but it doesn't track the inventories in work centers or stock locations. Kanbans are just tickets to authorize the movement or production of parts. Most computer systems are not able to track Kanban transactions, which means there is no way to link Kanbans with the perpetual inventory records stored in its data base.

It is true that a company can count the materials transferred by Kanbans and enter this information into the computer, but this would be cumbersome and liable to error. There is also the problem that production Kanbans don't track labor, which requires another parallel system.

A combination of a repetitive manufacturing system and a Kanban system would provide a solution to these problems. The repetitive system would keep track of all material transfers and stock values on the computer's data base. The Kanban would take care of moving materials between work centers when they are needed.

A repetitive manufacturing system can be made to wait for a Kanban to arrive at a preceding process before it issues the assigned rate of parts. The mechanics of doing this are simple: The stock location operator issues the parts requested by the Kanban and then transacts the rate on the computer with a repetitive command.

Figure 4.8 shows a joint repetitive and Kanban system. The process uses Kanbans A,

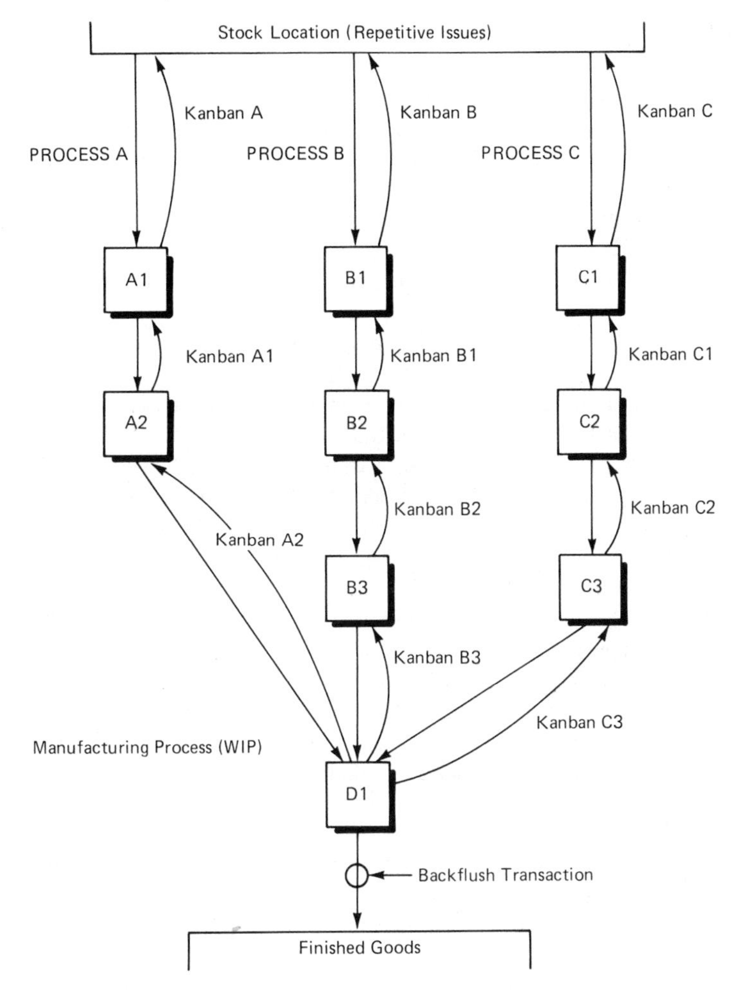

Figure 4.8 A repetitive-Kanban process combination.

B, and C to fetch materials from the stock locations. When one of the Kanbans arrives at a stock location, it will fetch the corresponding rate of parts called for by the repetitive schedule. Then the Kanban system will control all the movements of materials on the production floor without the need of executing repetitive transactions. In this arrangement, the process is considered to be a single inventory, a WIP location. After the product reaches the end of the line, a backflush transaction depletes the parts from the WIP location. Kanban movements on the manufacturing floor are transparent to the repetitive process. This Kanban-repetitive combination uses the best features of both systems to move and track parts only when they are consumed.

4.9 KANBAN TRACKING

Tracking Kanban status is one of the main problems that a materials planner is going to face in using such a system. In the work order and repetitive systems, the planner knows where the materials are and what has to be issued to meet a particular production schedule. In a Kanban system, the planner has global information about production levels and then uses this information to calculate the number of Kanbans required to support the schedule. But once the Kanban system starts to operate, the planner will lose control of the status of the Kanbans, for all the movements of material will occur at random, driven only by the Kanbans' usage rate.

The initial tendency of the planner is to make a log list with vertical entries for every Kanban in the system and to use it to track the status of every one of them. For a small Kanban system, this approach could work for a while, but as the number of Kanbans increases and the traffic becomes intense, the planner would spend considerable time doing nothing else but tracking. Just-In-Time considers this time a waste.

There are a few ways of reducing the additional workload in tracking Kanbans. For example, there is no point in keeping a Kanban uniquely associated with a particular part lot. All that need be known is the number of Kanbans issued for a part and the size of the lot they represent. A company would normally number Kanbans in sequence for convenience, but it shouldn't care which Kanban numbers are at any particular location. Let's assume there are three Kanbans of a particular part number in a stock location and the Kanbans represent a lot of twenty parts each. There is no need to know if they are Kanbans 3, 5, and 6 or Kanbans 1, 4, and 7. The effort in tracking them this way would be a waste. The only important information is that there are three Kanbans in that location with a total of sixty parts.

There is also a tendency to track the lead time of Kanbans. It is considered very important that Kanbans meet their estimated lead times, for this affects the supply of parts to subsequent processes.

One way to solve the tracking of Kanban lead times is to provide every work center with an estimate of how long it should take to process a Kanban. The operators are then asked to report any deviation from that estimate. Tracking would thus focus on default cases rather than normal cases.

A simple Kanban reporting system would require the operators at every work center to report at a particular time of day—preferably at the end of the shift—the number of Kanbans they have at hand and the number that are past due. This report probably will not take more than one page, and it will give the materials planner an overview of the status of the materials. It will also provide the opportunity to detect supply problems in advance if the number of Kanbans expected in a work center are not adequate to meet the production rate specified for the next shift.

4.10 SUMMARY

After reading this chapter, it should be clear that the implementation of a Kanban system requires numerous decisions. For example, how many part numbers should be put under a

Kanban initially? Should the system be implemented in house first or with suppliers? Should class A parts, class B parts, or subassemblies be picked?

It is difficult to answer such questions in a way that will satisfy all new practitioners at the same time, for every manufacturing organization and process is different, no matter how close the products. Nonetheless, here are some general guidelines. First, it is recommended that the implementation of a Kanban system begin in house rather than beginning with outside suppliers. The initial experience and the knowledge gained will prepare the materials department to deal with similar problems once the Kanban travels out of the factory.

The safest method is to pick a few work centers and design on paper a Kanban system for them. Put in writing the Kanban's operating procedures and design the Kanban card and the information it should carry. Make sure that those involved review the procedures and provide some useful feedback. Once the procedures are finalized, conduct a training class to make sure everyone understands how to operate under the Kanban system. Then select the locations of the Kanban mailboxes, the staging areas, and the material containers.

Once the system and training are in place, the materials planner can issue the Kanbans to the process. It is advisable that the number of Kanbans be on the conservative side. The planner must take into consideration the current status of the process, the yields, and the quality records of the parts.

The materials department should also determine the procedures for keeping track of materials and labor on the computer. It is best to have a repetitive manufacturing system in place before implementing a Kanban system.

Finally, the planner must put in place a simple tracking system before releasing the first Kanbans. If suppliers are involved, the planner must make sure that they understand how to handle the Kanbans so they don't get lost.

After the Kanban system is working for a few weeks, it is helpful to have a review meeting with all the parties involved to make the changes necessary to correct any problems. A few weeks later, start tuning the system and reducing the number of Kanbans by reducing the initial safety factors used. The system should soon start producing excellent results by controlling the transfer of materials in a true pull fashion.

One final word of advice: Do not use a Kanban system with a work order system. The systems are incompatible. The paperwork associated with work orders would complicate the Kanban system and slow the process down, producing very high levels of overhead and waste.

REFERENCES

BELT, BILL. "MRP and Kanban: A Possible Synergy?" *Production and Inventory Management* (American Production and Inventory Control Society) 18, no. 1 (1987): 71–80.

HALL, ROBERT. "Kawasaki U.S.A. Transferring Japanese Production Methods to the United States: A Case Study." (American Production and Inventory Control Society).

JAPAN MANAGEMENT ASSOCIATION, ed. *KANBAN: Just-In-Time at Toyota.* Translated by David J. Lu. Stamford, Conn.: Productivity Press, 1985.

MELNYK, STEVEN A., and RICHARD F. GONZALEZ. "MRP II: The Early Returns Are In." *Production and Inventory Management* (American Production and Inventory Control Society) 26, no. 1 (1985): 124–137.

MILLARD, ROBERT I. "MRP Is None of the Above." *Production and Inventory Management* (American Production and Inventory Control Society) 26, no. 1 (1985): 22–30.

MONDEN, YASUHIRO. *Toyota Production System.* Atlanta Norcross, GA: Industrial Engineering and Management Press, 1983.

SCHONBERGER, RICHARD J. "The Kanban System." Appendix in *Japanese Manufacturing Techniques: Nine Hidden Lessons in Simplicity.* New York: The Free Press, 1982.

SHINGO, SHIGEO. *Study of "Toyota" Production System from Industrial Engineering Viewpoint.* Tokyo, Japan: Japan Manufacturing Association, 1981.

5 ★

Time Waste and Just-In-Time

One of the major features of a Just-In-Time system is the crusade against waste. We normally associate waste with material things. In a factory, we say there is waste when the workers scrap material or when they invest labor in reworking parts to make them functional. But we very rarely talk about waste when material is waiting for a machine tool setup. We also ignore buffer inventories waiting for a process or the traveling time of parts going from one location to another. Just-In-Time considers time waste a major factor that consumes resources and produces no return.

In a factory, time waste is related to the labor required to build a product, for every product is associated with standard labor hours (i.e., the time it takes to put the product together). These labor hours determine not only the direct labor cost but also the overhead associated with the product. In general, the cost of every hour that a worker uses to assemble a product is magnified by the overhead rate of the department.

Just-In-Time broadens the concept of time waste to include more than the labor hours invested in building a product. Material traveling from one work center to another is a simple example. In a non-Just-In-Time factory, material travel time is less critical. The process has plenty of buffer inventories which relaxes the need for a fast delivery of materials to work centers. In a Just-In-Time system, material travel time is of more concern. There are no buffer inventories, and the worker depends on material from upstream processes to continue work. The material travels in smaller quantities more frequently; thus the time it takes to go from one work center to another must be minimized.

Another important area of time waste consists of the setup times for machinery. In Just-In-Time, a production lot has a small number of units, which presents the problem of frequent tooling setup changes needed to process different parts. The same problem arises when a multiple product production line switches from one job stream to another. The Japanese have mastered the reduction of setup time by means of the single-minute exchange of die (SMED). Shigeo Shingo created the SMED concept in Japan during the 1950s. Shingo's initial work concentrated on the reduction of the setup time in production to under ten minutes. This chapter reviews the principles of the SMED system and how it can be extended to other areas of manufacturing.

5.1 MATERIAL FLOW AND TIME WASTE

When a part arrives at the receiving dock of a manufacturing plant, it follows a determinate route in the factory. This route takes the part through the process until the line workers use it in assembling a product. Let's pick one part and stay with it all along the process. We will note periods of idleness and activity as the part advances through the process.

In a typical organization, the receiving department will receive a part and match it against an open purchasing order. Then the receiving clerk will send the part to the receiving inspection department. In this department, an inspector will check the part against a set of prints to make sure it conforms to its specifications. If the part passes inspection, the inspector will send it to the stockroom. In the stockroom, the part waits until the stock clerk pulls it into a kit and sends it to the manufacturing process. Then the process integrates the part into higher assemblies until the final product reaches the finished goods inventory.

The time the part is in the factory can be divided into three kinds: moving, waiting, and processing. Just-In-Time considers moving and waiting time to be wasteful because the part is doing nothing. Conversely, workers are adding value to the part during processing time. Under Just-In-Time, any activity that doesn't add any value to a part is a waste and must be minimized.

Moving

Moving is transportation. Workers move a part from one process to another or from the receiving dock to a stock location or a process in the production line. The part might even travel from one building to another. Moving does not add any value to the part.

There are several ways to reduce moving time. One is to reduce the distance parts have to travel by improving the flow of the process. Manufacturing and industrial engineers can design the layout of the factory to minimize the distance the parts have to travel. Such layouts are usually process-oriented. In some instances, it has to be a compromise so that the majority of parts travel an average minimum distance.

Another way to reduce moving time is to speed up the delivery process. This can be done by using a material-handling system (see the next section). There is a very close relationship between the factory process and the type of material-handling system used to move the parts.

The physical size of the factory is another factor that influences the moving time of parts. Just-In-Time calls for the elimination of buffer inventories and the use of small production lots. This results in less need for storage space on the manufacturing floor. The factory can be compacted and processes can be placed close to each other. The smaller distances will reduce the time needed for parts to travel. It will also reduce rent and utility expenses.

A side benefit of reducing the size of a factory is the improvement in communication among the workers, increasing their efficiency in exposing and solving problems.

In summary, there is no value added to a part during movement. This activity is a waste and must be minimized.

Waiting

Waiting occurs when parts are stored in stockrooms or in buffer locations on the manufacturing floor. Parts in a stockroom are waiting to be used and therefore are wasting time and money.

From the process point of view, parts waiting are not gaining any value. There is no manufacturing activity associated with the part during that time. In reality, the parts are wasting not only time but money. The manufacturer is losing the interest that it could have made by investing the money spent on those parts differently. The manufacturer is also wasting the rent and utilities for the space used to store the parts.

Times of idleness are not always recognized as such. Everyone realizes that parts in a stockroom or buffer location are waiting. However, parts on the factory floor can also be waiting during the setup of a machine tool. In general, these parts are considered to be in the production process, but in reality they are waiting and wasting time. The SMED system directs its effort to reduce machine setup times to optimize production runs (see Section 5.3).

Storing parts in buffer locations along the production line is a common practice in most manufacturing organizations. One cause of buffers is issuing parts to the manufacturing floor at a higher rate than the consumption rate of the production process. Suppose workers are processing 100 sets of parts per day but the stock clerks issue kits of 300 parts at a time. That creates a buffer of 200 parts the first day and a buffer of 100 parts the second day. By the third day, the clerks will probably issue another kit of 300 parts, repeating the process again.

A repetitive manufacturing system is excellent at eliminating line buffers. Under this system, the production consumption rate is matched with a continuous stream of parts. Then a feedback mechanism is used to tune the stream to the variations in rate so that there is never excess material in the line. This eliminates the problem of having parts waiting for a process. A material-handling system designed to move parts smoothly and without buffer locations is a good companion for a repetitive system.

For smaller companies without the funds to invest in a repetitive system and in a material-handling system, a Kanban system would suffice. Parts will only move when the downstream process consumes the parts at hand.

Here is a simple exercise that might well be worth the trouble. Ask your manufacturing engineers to walk along the production line, following the process flow. Every time the engineers see a group of parts not active in the process, they should make a note. Later have them come back to those points and study the process to understand what is causing the parts to remain idle. The goal is to work on the process until the idle time of parts and the size of buffers are reduced to a minimum. This goal might require several iterations in order to fine-tune the balance of the production line. It also might require improving the synchronization of different processes or reducing the setup times of several machines. At the end of this exercise, however, the process will be leaner and there will be few buffer parts.

Processing

Processing is the only activity that adds value to a part. The production process adds value to a part by applying direct labor or machine labor. Just-In-Time considers this activity necessary, but it always tries to minimize its length.

Reducing production time can be done most effectively during the design process. Later, after the product is in production and has moved up the learning curve, there will always be room for reductions in production time.

Section 5.4 reviews some basic ideas for reducing production time. Reduction will become more important as the level of automation increases in modern factories, for it involves not only human labor but also machine labor. In a Just-In-Time system, there must be a never-ending effort to reduce the time it takes to produce a product. This effort will be paid back many times over by increasing the productivity of the system.

Figure 5.1 shows a simple breakup of the three states of a part as it goes through a two-step process. The stockroom is usually the place where the part is idle most of the time. Moving time is the shortest but moving also occurs most frequently for a multistep process.

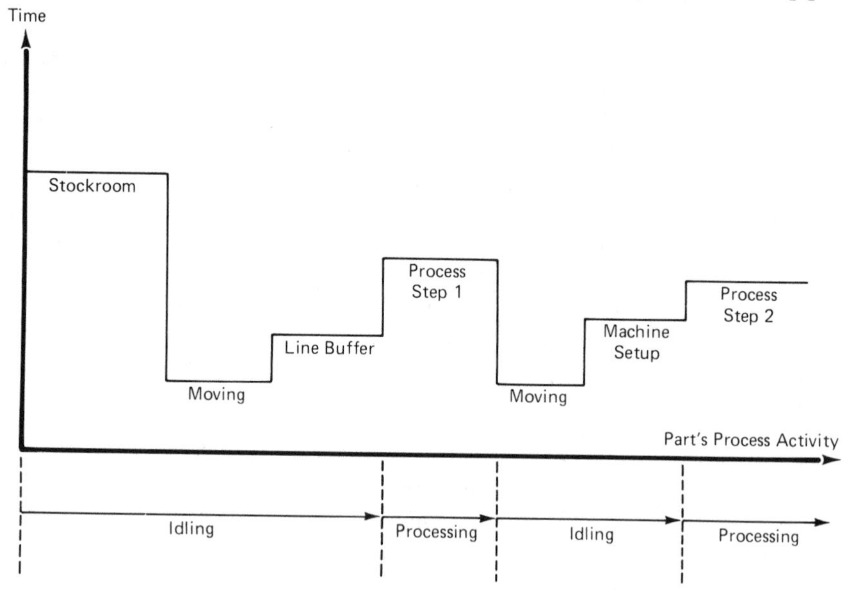

Figure 5.1 Simple analysis of the status of a part during the production process.

The efficiency of parts in the process can be measured by calculating the ratio of idle time to process time. For example, an 80-percent processing efficiency would mean that the parts are idle only 20 percent of their process cycle time.

5.2 MATERIAL-HANDLING SYSTEMS

A Just-In-Time system demands the prompt delivery of parts to the work centers in the exact quantities required. This means that the right part must go to the right work center at the right time. Just-In-Time also calls for frequent deliveries of small quantities of parts and a minimum buffer inventory between processes. The material-handling system must obviously play an important role in the overall implementation strategy for a Just-In-Time system.

In general, the automation of a factory depends on the process volume, type of product, and cash resources. One of the key aspects of automation is the design of a material-handling system to move the parts in the process. Small companies with limited cash resources will rely more on a manual material-handling system than will large companies with high-volume processes and financial muscle. A company of the latter kind will probably design an elaborate material-handling system which will require little or no intervention by workers. In any event, the implementation of a material-handling system is necessary for a successful Just-In-Time system. This doesn't mean that a company has to spend a lot of money for a sophisticated system. It simply means that it has to give considerable thought to the design of a system adequate to its needs.

Material-Handling Strategy and Goals

Before implementing a material-handling system, a company needs to define the strategy to use in developing such a system. First, it must define the system's goals and how they relate to the Just-In-Time philosophy. Once it determines the goals, it needs to consider the practical factors related to the type of product, the production rates, and the cash investment it is willing to commit. In any event, the material-handling system must handle materials in the factory in accordance with the Just-In-Time rules.

There is a set of basic goals a material-handling system must achieve at the point of implementation. These goals are not necessarily cast in concrete, but they generally represent present and future requirements.

The set of goals include the following:

- The system must have adequate capacity to handle the volume requirements of the production line.
- The system must be flexible enough to handle future products assembled in the factory.
- The system must be able to expand to meet future capacity increases.
- The system must be able to store production materials for one day's production or less.
- The system must be reliable and subject to a minimum of downtime.

Manufacturing and industrial engineers are usually responsible for designing a material-handling system. They have at their disposition many different ways to implement such a system. In general, the final implementation will depend on the type of product and the volume the factory has to process. For example, a material-handling system for a factory that builds four hundred computer systems per month would be very different from that of a factory that builds fifteen thousand disk drives per day. They have different volume requirements but both might have the same basic goal of supporting a Just-In-Time system. Figure 5.2 shows a material-handling system for a company with medium-volume capacity.

Capacity. Type of product and capacity are the first two factors the engineers need to determine before they start designing a material-handling system. A factory with a small product and a large volume will require a different approach than one with a bigger product and a lower volume.

Determining the type of product is usually simple, for companies tend to have well-defined product lines. Determining capacity confronts the problem of growth. Company management must decide how much growth capacity they want to invest in. They certainly wouldn't want to invest in a system that couldn't handle the factory's growth in the near future. But they also wouldn't want to spend money providing for capacity that will not be required until five years down the line. In each of these cases there would be a waste of resources and a violation of Just-In-Time concepts.

The spare capacity goals of the material-handling system should depend on the growth rate of the company. A company experiencing relatively flat growth should be more conservative than one experiencing explosive growth.

Flexibility. Once installed, material-handling systems are inflexible unless the engineers design them with flexibility in mind. Under Just-In-Time, the problem of flexibility becomes accentuated, because the system will constantly demand process improvements and changes in the way workers do their jobs.

In general, a company should design enough flexibility into its material-handling system to allow adaptation to process and product variations. This is not easy to do, because such systems involve a lot of hardware. For example, a conveyor system is very efficient in transporting material from one work center to another, but it is very inflexible regarding production flow changes.

One mistake to avoid is to tailor the material-handling system to the current process and products. The engineers designing the material-handling system should try to make the system as independent as possible from the products the system will transport. Such independence will pay handsomely in the future, when current products evolve or new ones go into production.

Designing for independence brings a certain degree of risk. The company must be assured that the material-handling system will be able to handle the current products at the

Figure 5.2 Material-handling system at Metaphor Computer Systems.

specified volumes. It would be catastrophic if the system, once installed, couldn't handle the requisite volume.

To conclude, flexibility in product and process variations is critical to consider at design time. Before a new system is approved, proof must be given that it will be capable of handling the present processes in the factory. The system and the processes must work together efficiently without compromising the goals of flexibility, capacity, and performance.

Delivery and Storage of Parts

The job of a material-handling system is to deliver parts to work centers efficiently and with a minimum of worker intervention. The designers of a material-handling system must plan the storage level for parts the system is going to allow during normal operation. Of course, a good material-handling system will generally deliver parts on time, but in a real environment small variations in the process and work centers may well cause delays. Process buffers are usually the solution. Since Just-In-Time calls for the reduction of all buffers, so a compromise must be found.

In general, a company needs workers to have parts available to do their jobs at the work centers. A worker should never be idle because of lack of parts. On the other hand, a worker shouldn't produce more parts than are required to meet the demand of the downstream process. A simple solution to this problem is to allow storage of parts for a particular period of time. For example, the company might allow the material-handling system to store parts for one day's production only. For high-volume processes, the rule might be to allow only a few hours' supply.

The physical restrictions can be used to determine the parts stored in a material-handling system. For example, a simple material-handling system might consist just of workers moving parts from one work center to another by using metal carts and storage shelves at both ends. In this example, limiting the number of carts and shelves is a way to control the amount of products the system can handle. A company could calculate the number of parts required for a day's production and divide it by the number of parts a shelf can store. Then it could check the flow of parts in the process and calculate the number of carts required to transport them. These two calculations could be used to prevent the system from transporting or storing more parts than required.

Following this set of rules will help control material transfers and storage in a material-handling system:

- Decide on the maximum rate of parts the system can move and store before designing the material-handling system.
- Provide enough moving capacity and storage space to accept only that quantity of parts.
- Eliminate additional storage space that could lead to accepting unnecessary material (e.g., shelves, tables, and other flat surfaces, including floor space).

- Avoid the use of systems that lend themselves to becoming storage vehicles, for example, conveyors or carousel systems. They could easily become rotating stockrooms with no control.
- Provide very small buffer inventories on either side of a work center. The goal is to provide only enough cushion to cover for small variations of speed in the process. If there is need for more than that, something is wrong with the process. Go back and review the balance of the line.

A Simple Pegging System for Moving Material

This section presents an example of a pegging system to move material from one work center to another. The system, which illustrates the Just-In-Time principles, is very simple, but it can be expanded to fit any level of production or complexity.

Figure 5.3 shows a process with two work centers, A and B. Buffer stock locations A and B feed parts to the corresponding work centers. Buffer C receives the output from work center B. Let's assume that the process rate is four units per hour and that the buffers are designed so they will store only one hour's worth of parts. Also, let's assume the product is a cabinet. Workers will store the cabinets on the specific buffer areas on the floor. They will not be allowed to store parts outside the designated areas or to store more than four cabinets at a time. If at any time a buffer is full, the worker will have to stop producing until there is a space available.

When the line starts for the first time, worker B has no material in his input buffer B. The worker has to wait until work center A produces the first part and stores it in the buffer. During that time, Worker B can be assigned to another work center (if the idle time is long enough) or simply can wait. Once worker A delivers the first part to buffer B, then worker B will fetch the part and start processing.

Several things could happen during the normal operation of the system. For example, work center A's output rate could be higher than the consumption rate of work center B.

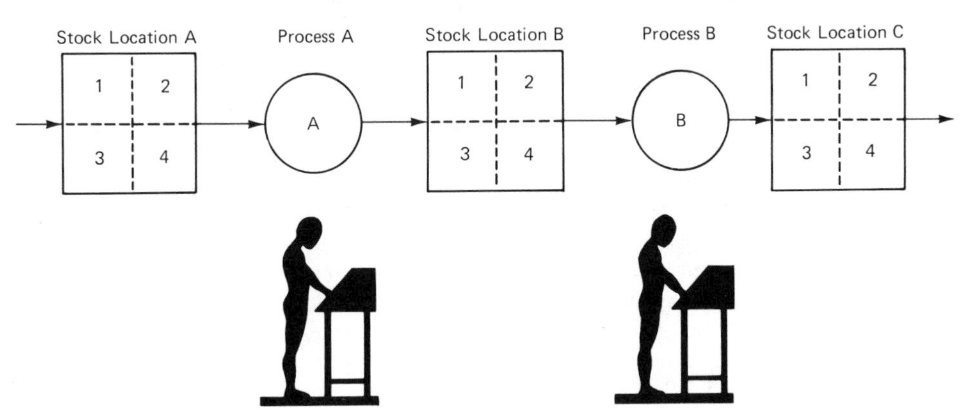

Figure 5.3 Two-step process using a manual pegging system.

Since worker A is not allowed to store parts in buffer B when it is full, this causes the worker to become idle until worker B empties at least one buffer slot. The operating mode will show any line imbalances immediately when the idle and busy times of the work centers are checked. A good line balance should maximize workers' busy times and reduce their idle times.

Conversely, worker B might consume parts faster than the output rate of work center A. In this case, buffer B will be empty most of the time, forcing work center B into idleness.

One way to correct line balances is to review the process operations assigned to every work center. This effort most likely will show that the load in one work center is higher than in the others, causing different rates in output. Manufacturing engineers could correct this problem by simply reassigning process steps to the work centers.

Another good feature of the system is that it responds quickly to process problems and hence minimizes any rework required. For example, let's assume that worker B finds a defect in the parts received from work center A. Worker B will stop processing immediately, resulting in the saturation of buffer B. A full buffer B will force worker A to stop production, for there is no more room to store output. The system has limited the rework to the size of buffer B.

Finally, let's assume the process is running smoothly, the line balance is ideal, and quality is at its highest level. It might be worthwhile to have the manufacturing engineers monitor the buffer levels at locations A, B, and C. Such monitoring could result in a reduction of the materials stored in these buffers. For example, reducing the buffers from four to two parts would result in a WIP reduction of 50 percent.

In a Just-In-Time system, one constant goal is to reduce buffer inventories, for the optimum buffer size is one unit.

A Mini-Cartrac System

One of the problems with the pegging system described above is that it requires firm discipline on the part of the workers. The buffer locations are square sections painted on the floor and do not physically restrict where workers can store excess parts. This section describes the implementation of an automated system that uses pegging principles but in a more forceful way. The material-handling system uses a Mini-Cartrac system built by SI Handling Systems and installed at the Metaphor Computer Systems factory.

Figure 5.4 shows the Mini-Cartrac main components. They are modular, which permits configuring the system to meet the needs of different manufacturing processes. Mini-Cartrac is a transport system for small products that uses computer-controlled nonsynchronous carriers. The system has controlled acceleration and deceleration capabilities to move the carriers gently. It also positions the carriers automatically within close tolerances for interfacing with the process. The system is intended for use in manufacturing plants, warehouses, and distribution centers.

Figure 5.5 shows a Metaphor Mini-Cartrac work center, consisting of modular spurs capable of queuing and transferring carriers with the assembled products. Metaphor's

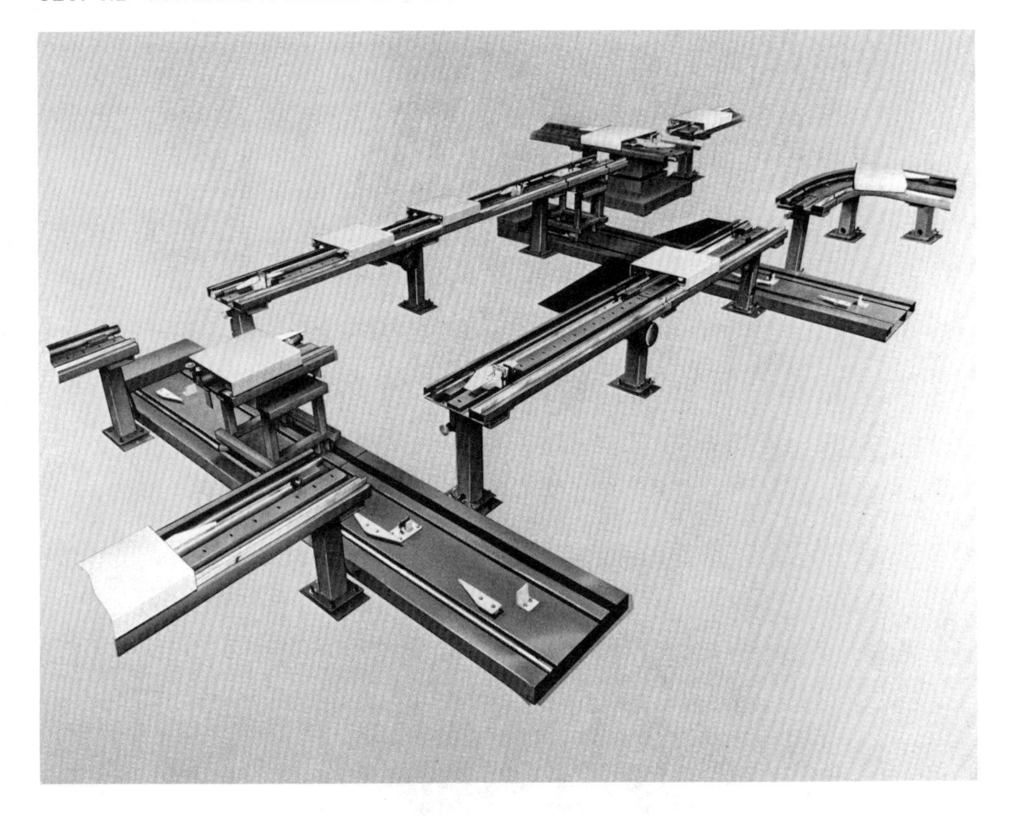

Figure 5.4 Main modular components of the Mini-Cartrac system.

products include workstations, file servers with hard disk pods and streamer tapes, communication servers, and relational data base servers. Metaphor uses the Mini-Cartrac to assemble workstations, keyboards, file servers, disk pods, streamers, and communication servers. All these devices are under 175 pounds, the maximum weight the Mini-Cartrac can transport. The relational data base server is too big and heavy for the system. At the work centers, the workers assemble the products on carriers, shuttling them to the next work center without moving from their places of work.

Once a worker finishes an assembly operation, he or she inputs the next work center's address into a keypad within easy reach. The system computer releases the carrier with the assembly into a queue station and dispatches the transfer car to pick it up. The transfer car then delivers the carrier to the next work center. The computer also releases into the worker's work center a new carrier with the next assembly. Workers waste no time moving material around or receiving the next job.

Metaphor's engineers designed the Mini-Cartrac layout to have one buffer slot in every work center. The system operates in true Just-In-Time pull fashion. No worker can

Figure 5.5 Mini-Cartrac work center at Metaphor Computer Systems.

send an assembly to the next downstream work center unless there is an empty slot available. This concept forced Metaphor's manufacturing engineers to balance the line to reduce bottlenecks in the process. Also, WIP is minimal because of the lack of buffer areas.

There are a couple of rules the Metaphor line follows in order to add predictability to the flow in the process. The first one is to do all troubleshooting off-line. Whenever a submodule doesn't work, the workers replace it with a good one. Then they send the faulty module to an off-line repair station. The second rule is that all electronic tests done in the line are of a Go/No Go nature. If the product passes the test, it will advance to the next downstream work center. If it fails, the worker will replace the faulty part for a good one and send the bad one off-line.

The other benefit of the system is that it forces the line to solve problems immediately. There is no room for storage if some work center discovers a problem in the process. Manufacturing engineers and support engineers are on call to address any problem quickly.

Metaphor has used the system for four years, and so far it has met all the initial expectations. Figure 5.6 shows the material dispatch center, which faces the production line. The monitor on the left tracks the transfer transactions from one work center to another. The monitor on the right records all material transfers on Metaphor's manufacturing computer. In the upper right-hand section of the picture can be seen the central transfer track that runs

Figure 5.6 Material dispatch center at Metaphor Computer Systems.

along the manufacturing floor. Figure 5.7 shows a transfer car shuttling a Metaphor workstation along the central transfer track.

The Mini-Cartrac system was ideal for Metaphor's products and production rates; however, other alternatives might be more suitable for different products and high-volume applications. For example, a high-volume manufacturer of hard disks would be better off utilizing a conveyor system to move the product in the production line rather than a Mini-

Figure 5.7 Transfer car shuttling a workstation along central track.

Cartrac system. The best material-handling system for a factory depends greatly on the type of product and the volume, and no one system will meet the needs of every manufacturer.

5.3 MACHINE SETUP TIME AND THE SMED SYSTEM

Just-In-Time calls for small lots and frequent production runs. This operation mode helps to control excess materials in the process, but it creates the problem of wasting additional setup time for machines. In a normal operating environment, the time wasted in machine setup becomes more evident when frequent small lots are processed. The single-minute exchange of die (SMED) system is a collection of techniques used to reduce machine setup time. As the Just-In-Time idea of using small lots evolved, SMED became associated with the system. Soon SMED became broadly accepted as a way to reduce setup times both in heavy industry and in light industry. This section explains how to use SMED concepts to set up entire processes to build different products using the same production line. The result is an efficient utilization of the capital invested in material handling and work centers.

The SMED System

Shigeo Shingo developed the SMED system in Japan during the fifties and sixties. At the time, Shingo frequently visited several heavy industry companies, including Mazda and Toyota, with the purpose of helping them to improve their productivity. The problem the companies were facing was that it took a long time to execute setup changes in their heavy machinery. Shops were measuring setup times in hours, and the shop managers were under pressure from management to reduce them. It was not until years later, after Toyota embraced Just-In-Time companywide, that SMED became an integral part of the Toyota production system.

The SMED system is a process of systematic machine setup analysis that clearly distinguishes every step in order to introduce timesaving changes. The goal of SMED is to increase the productivity of machines by reducing their idle time and to reduce machine setup from hours to minutes.

Setup Time Analysis

The first step in applying SMED to a particular machine is to analyze the setup time for that machine. This analysis must clearly identify two types of setup. The internal setup requires the machine to stop operating. During this setup, the machine is not productive. The other is the external setup, which can be done with the machine operating.

The distinction between internal and external setup will yield an immediate productivity improvement of 30 to 40 percent. After this step, the goal becomes to convert internal setup tasks into external ones. The conversion of setup tasks is an iterative process. The final step is to reduce the time required for the tasks by using new production methods.

After many iterations, SMED will increase the productive time of a machine and reduce the idle time required for a new setup. The following three steps constitute the reduction process.

Step 1: Identify internal and external setup tasks. Figure 5.8 analyzes the setting up of a machine to do a new task. The setup time is normally linear and the machine is not productive during the conversion. The first step in applying SMED is to study all the different subtasks involved in setting up the machine. The next step is to group them in two categories: internal and external. Then the operator changes the procedure for setting up the machine, executing all the external setup tasks while the machine continues to operate. Typical external setup tasks include moving tools, dies, setup jigs, and raw materials for the machine's next job.

Step 2: Change internal setup tasks into external setup tasks. Figure 5.9 shows the improvement achieved by changing some of the internal setup tasks into external ones. The changes reduce the idle time of the machine and increase productivity. This is because external tasks, unlike internal tasks, can be executed while the machine is still operating.

Examples of external tasks include preheating dies before installing them, preparing cutting dies on separate benches before installing them, and loading new material into the machine while it is running.

Step 3: Reduce the setup times of all operations. Once setup tasks have been changed from internal to external tasks, the next step is to work on reducing the time it takes

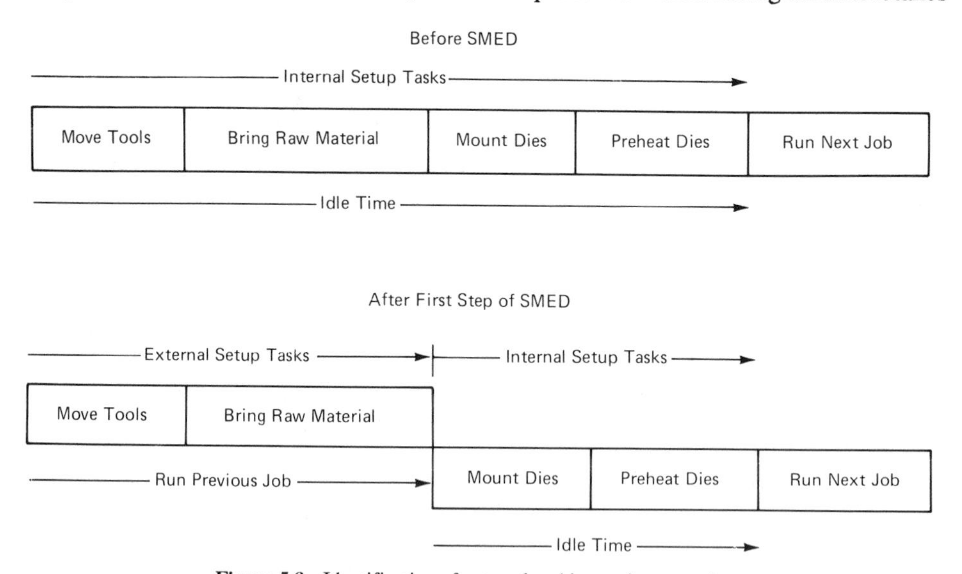

Figure 5.8 Identification of external and internal setup tasks.

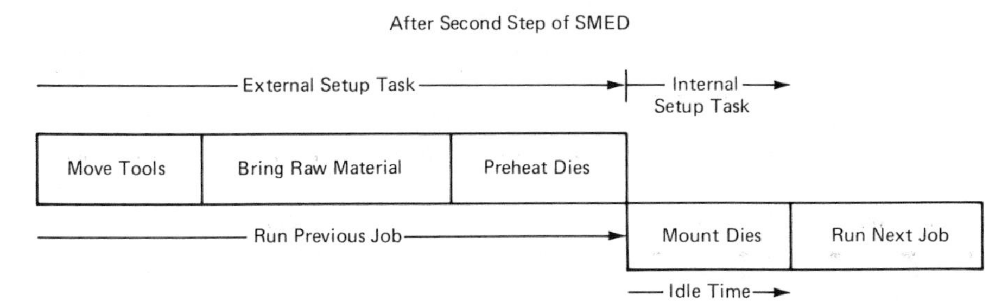

Figure 5.9 Change of internal setup tasks into external ones.

to complete all the tasks. This might require that engineering changes be made in the tools or machines. It might also require the design of new jigs. The thrust of the effort is to reduce the labor required to complete the tasks. Figure 5.10 shows how this step increases the operating time of the machine by reducing the idle time for setup.

Methods of reducing setup time include improving die positioning methods, improving material feed methods, improving hose connections and hookings, standardizing die dimensions and positioning, eliminating fine adjustments, standardizing bolts, using color marking to improve efficiency, improving operator training procedures and communication with other related departments, and preparing teams to work on problems as they occur.

SMED and Related Processes

One natural extension of the SMED system is to apply its concepts to other areas of the manufacturing process. But before this can be done, the concepts of internal and external tasks need to be broadened. A task is to be considered internal if the worker's labor adds value to the product (e.g., the labor used in assembling a product). Conversely, a task is to be considered external if the worker's labor adds no value to the product (e.g., loading materials on the work center's shelves).

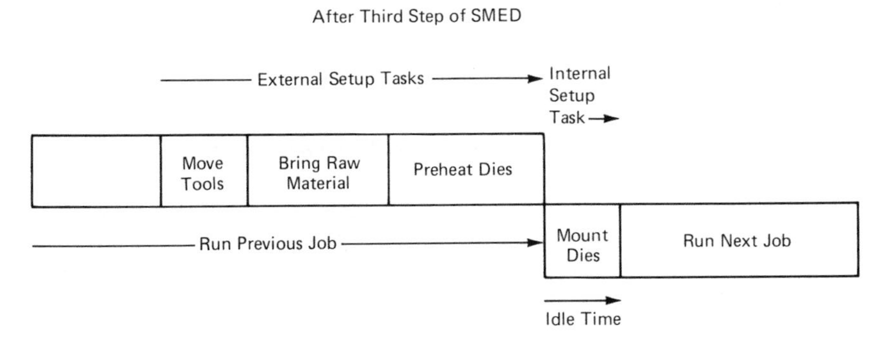

Figure 5.10 Reduction of setup times of all operations.

Using these definitions, a company can execute a methodical analysis of its production process, categorizing the tasks as either external or internal. The manufacturing engineers can then follow the three steps presented in the previous section. The goal is to make as many tasks external as possible and then to reduce the time needed to do them. Below are several examples to illustrate this process.

Electronic testing of printed circuit modules. Electronic testing of printed circuit modules is a complex task that requires sophisticated equipment and trained workers. Testing modules is normally a two-step process. The first step is to test a module in a coarse way using automatic test equipment. This equipment is expensive and in general does not guarantee that the electronic module is 100 percent functional. The second step, commonly called functional testing, requires testing the module with test equipment that closely simulates the functional environment of the final product. This task entails that the worker has to plug the module into connectors and cables.

Functional testing could take a few minutes, and usually it takes as much time to plug in the modules as it does to perform the test itself. In addition, problems sometimes arise because the cables and connectors start to wear from use, causing failures and false results.

In applying SMED to this process, the task of plugging the module into the tester would be classified as external. The actual testing would be classified as internal. The next step is devise a way of performing the external task during the performance of the internal one. The final step is to reduce the time it takes to execute both tasks.

To overlap the tasks, there has to be a way of mounting the next module while the current one is under test. Figure 5.11 shows a way of doing this—by adding another mounting fixture to the tester. The worker will mount the next module on this fixture while the tester is testing the current one.

The final step is to reduce the time it takes to execute the tasks. Let's assume that engineers have already optimized the test programs. Therefore, the only area for improvement is the mounting of the module to the tester. The mounting can be done by using vacuum fixtures. Having the test electronics of a functional tester interface a vacuum fixture can reduce the plugging time of the modules substantially. The system will also be much more reliable than the conventional method of plugging into connectors and cables. The operator of the tester will simply mount the module using aligning pins and press a button that activates a vacuum valve. The board will then be sucked into position. The worker will wait until the current module finishes testing and press the starting button again for the next test.

Reducing material loading time in a work center. A production line in operation requires a constant supply of parts to keep workers productive. This operation starts in stockrooms and buffer locations and ends when the materials arrive at the work centers. For an assembly operation, the work center operator most likely will store the parts on shelves or containers, readily within reach. The stockroom clerk normally pulls parts in parallel with the work center, avoiding any delays in the process. The time it takes a worker to unload parts in the work center directly affects his or her output.

Applying the SMED concepts, let's call the task of loading and sending parts to the

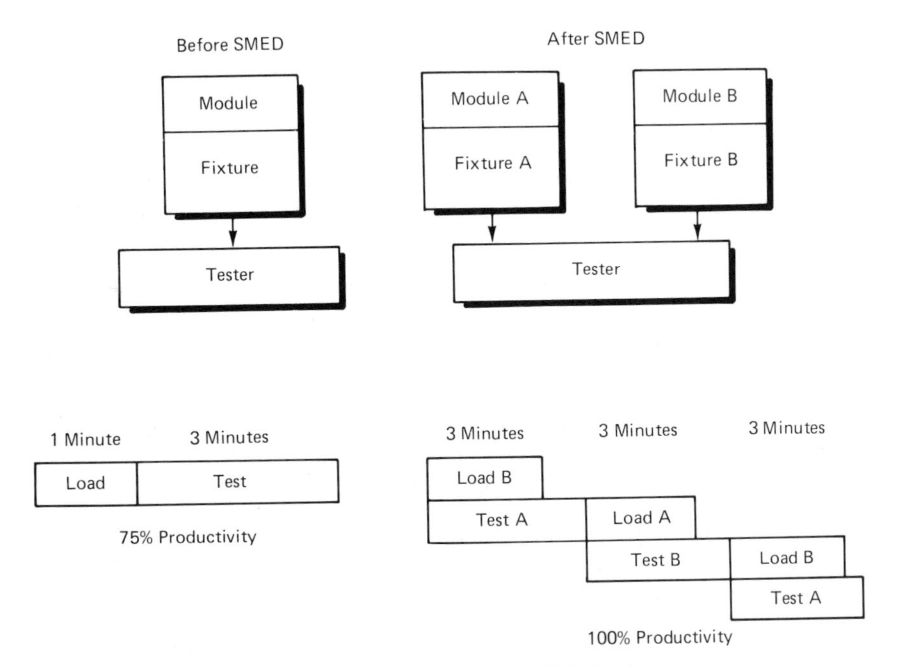

Figure 5.11 Electronic testing using SMED techniques.

work centers an external task. Ideally, this task should be performed in parallel with the internal task of building the product. However, imagine there is only one worker at each work center, so the workers have to stop producing to unload the parts.

For example, let's assume that a worker at a work center receives batches of 50 sets of parts stored in large containers. The worker unloads these parts and stores them in small boxes attached to the work center's shelves. A simple way of making this task external is to use bin boxes that the worker can remove easily. Then a system using two sets of bin boxes can be instituted. One set is normally at the work center with the parts needed to assemble the products. The other set of bin boxes is in the stockroom, where the stock clerk fills them with the next set of parts. When the set at the work center is nearly empty, the clerk will send the set from the stockroom. The worker at the work center will simply replace the empty boxes with the full ones and send back the empty boxes to the stockroom for refilling. This method is much faster than using one set and having to unload the parts and store them on shelves.

5.4 PROCESSING TIME

At the beginning of this chapter, it was noted that a part in a factory spends time moving, waiting, and being processed. The Just-In-Time system considers moving and waiting wasteful, for they do not add any value to the product.

Just-In-Time does not accept processing time as a constant, and it tries to reduce it as much as the other kinds of time. Reductions in processing time increase the productivity of the factory.

Below is a set of rules for analyzing and reducing processing time.

1. Establish a processing time goal at the beginning of the product design cycle. This is the best point at which to determine how long it should take to build a product. Change the design to meet the goal if necessary.

2. During the design cycle, keep in mind that a product must meet not only assembly times goals but also disassembly time goals. For example, there is sometimes a need to disassemble a product to do line and field repairs. If a product is very difficult to disassemble, it will add time to the process.

3. Define the process clearly before engineering releases the product to manufacturing. Make sure that the process documentation matches the engineering documentation. If not, change the engineering documentation to reflect the process.

4. Make sure that the manufacturing engineers have clear procedures that cover the process step by step at every work center. The engineers must also train all workers involved in the process until they fully understand the tasks at their work center.

5. Define the quality system of the process. Make sure the workers understand their work center's quality goals and yields. Have a tracking system to monitor the quality levels and yields of the product along the process.

6. Develop a system to feed materials to the process. This system should include pull procedures for the stock locations, procedures for transportation to work centers, and unloading procedures at the work centers.

7. Once the production line is working, ask the manufacturing engineers to time every operation as part of a line-balancing reassessment. Tune the process to improve the line balance.

8. Make a list of the different tasks assigned to every work center, ranking them by length of time. Ask the manufacturing engineers to work together with the design engineers and the industrial engineers to reduce the task times. Do not confuse these reductions with the learning curve, which will produce results as the volume of the product increases.

9. Never stop questioning the time it takes to complete every operation. There is always room to do a task faster and better.

5.5 SUMMARY

Time waste is a very important subject, for it affects the productivity of every manufacturing organization. In a Just-In-Time system, a serious effort is made to reduce not only moving and waiting time but also processing time. To succeed, the effort should be systematic.

It is very important to study carefully how time is spent in the production process. Then priorities should be set regarding the areas to concentrate on. Reducing time waste is a never-ending process.

In some instances, it will take a major investment in capital equipment to substantially reduce time waste. (For example, it might require the procurement of a material-handling system.) Hindrances to investing in such equipment should not be taken as an excuse to avoid improving the way the workers transport materials. A simple pegging system needs only a marked area on the floor to produce some results.

In practice, the reduction of waiting time is related to the reduction of inventory. In a manufacturing organization, the stockroom is where most materials remain idle.

Another way of saving time is to make the factory more compact. Flat surfaces have a tendency to get loaded with materials, some of which eventually become obsolete. It is best not to allow floor storage and shelves in the line to store materials for more than one day's production. That is all the workers need to do their jobs. This also entails there will be daily schedules and linear production runs.

It is important to be firm about never issuing incomplete sets of parts to the work centers. If there is a line shortage, do not issue the parts and then stop the line. Nothing helps more to solve shortage problems than to follow this rule.

Just-In-Time calls for a relentless effort to reduce processing time. Every minute of processing time saved reduces the direct cost of the product. Waiting for the learning curve to reduce processing time is ineffective. Instead, give managers the goal of reducing processing time and never accept the claim that it cannot be reduced any further. Also, manufacturing engineers should constantly feed back process information to design engineers. This will help manufacturing and also make the next product easier to build. The goal is to design products that require a minimum amount of moving, waiting, and processing time.

Finally, always remember that moving and waiting time affects overhead rates and that processing time affects direct cost. All three kinds of time affect the total cost of a product. There must be always efforts to reduce them, and such efforts will always pay off in the long run.

REFERENCES

HARTLEY, JOHN. *FMS at Work*. Bedford, UK, IFS (Publications) Ltd. and North Holland, 1984.

HERNANDEZ, ARNALDO. "Just-In-Time for a Start-up Company." Paper presented at the Society of Manufacturing Engineers EMTAS 1986 Conference, Dearborn, Mi., March 1986.

HOLLIER, R. H. "Automated Materials Handling." In *Proceedings of the Third International Conference, Birmingham, 1986*. Bedford, UK, IFS (Publications) Ltd., 1986.

KOCHAN, ANNA and DEREK COWAN. *Implementing CIM*. Bedford, UK, IFS (Publications) Ltd., 1986.

MONDEN, YASUHIRO. *Toyota Production System: Practical Approach to Production Management*. Atlanta Norcross, GA, Industrial Engineering and Management Press, 1983.

MULLER, WILLI. *Integrated Materials Handling*. Bedford, UK, IFS (Publications) Ltd., 1985.

RANKY, PAUL G. *Computer Integrated Manufacturing*. London, UK. Prentice-Hall International, 1986.

SHINGO, SHIGEO. *Study of "Toyota" Production System from Industrial Engineering Viewpoint.* Tokyo, Japan, Japan Management Association, 1981.

———. *A Revolution in Manufacturing: The SMED System.* Stamford, Ct., Productivity Press, 1985.

TOMPKINS, JAMES A. "Material Handling and Storage: Making Hybrid Systems Work." *CIM Review: The Journal of Computer-Integrated Manufacturing Management* 1 (Spring 1985): 13–15.

Just-In-Time and Suppliers

A critical aspect of the Just-In-Time system is the development of close relationships with suppliers. Just-In-Time calls for fewer suppliers delivering high-quality products in small quantities and on time. A supplier working in this framework will require a different kind of relationship from the traditional adversarial relationship so common in the business world. In this new partnership, a manufacturing company will make long-term commitments to suppliers and provide them with accurate forecasts and technical support.

Some suppliers will resist the changes that a Just-In-Time system requires. To overcome this resistance, a manufacturer must put in place an education program for suppliers early in the program. Just-In-Time will force suppliers to change the way they conduct their manufacturing and shipping operations. It will also force them to reach higher product-quality levels. All of this requires additional effort with no immediate return. A manufacturer needs to show suppliers that in the long run they will also benefit from Just-In-Time. It will make them more competitive and create a loyal clientele.

Another aspect of a Just-In-Time supplier program is to reduce the number of suppliers that provide parts for the products. The goal is to have a closer relationship with a smaller set of suppliers. This is contrary to the idea of second source and competitive bidding. Also, it makes the task of selecting Just-In-Time suppliers more critical. For once the process starts, a manufacturer will not have a buffer inventory to cover in case of delivery and quality problems. The manufacturer's fate will depend heavily on the performance of the suppliers.

The main goal in developing a team of Just-In-Time suppliers is to create partnerships. A manufacturer should try to nourish long-term relationships where information and support flows in both directions. If a supplier does not understand or sympathize with this goal, then the company should select another supplier, one willing to cooperate. The target is not 95 or 98 percent performance, but 100 percent. With hard work, good communication, and good planning, this goal can be achieved. That's what Just-In-Time is all about.

6.1 A JUST-IN-TIME SUPPLIER PROGRAM

A Just-In-Time supplier program requires careful planning and execution. To avoid resistance to the program, a manufacturer must stress the concept of partnership from the beginning. The goal is to institute a system that simplifies the supply of parts to the factory and brings benefit to both partners. In other words, it should be a win-win relationship.

Once the supplier is in the Just-In-Time program, the manufacturer can minimize its inventories and eliminate any receiving inspection on the parts. These operational changes require the supplier to provide parts of excellent quality and to make consistent deliveries. To ensure success, the manufacturer should implement a total quality control (TQC) program in parallel with the Just-In-Time program. Chapter 8 outlines the fundamentals of a TQC program. It is necessary to have a TQC program in place before changing to frequent deliveries and short lead times.

Release Schedule and Delivery Rates

A Just-In-Time program with a supplier calls for steady and frequent delivery of parts. For example, in cases where the supplier now delivers parts on a monthly basis, the program would call for weekly or daily deliveries. For a high-volume production line, the supplier may be asked to make hourly or daily deliveries. The idea is to tune the delivery of parts to their consumption rates in the process.

Normally, in an MRP production environment, the MRP explodes the master schedule down to the requirements for every part. The MRP takes in consideration the materials at hand, the factory's lead times, and the suppliers' lead times before it prints the parts order point dates and quantities. Then buyers order those parts by issuing purchase orders to suppliers. Weeks or months later, the suppliers deliver the parts without regard to the actual consumption rate in the factory at that time. This activity builds inventory and consequently wastes time and money.

Just-In-Time calls for a reduction in a supplier's lead times. This task allows the manufacturer to tune the supplier's material releases to the actual consumption rates in its factory. Then it reduces the factory's lead time in order to obtain a substantial inventory reduction.

The supplier's release and schedule process can be divided into five steps. (The assumption is that the supplier has a lead time reduction program in place and can thus respond to the following rules.)

1. Make a long-term purchase commitment to the supplier. The length of the contract should be roughly eighteen to twenty-four months. This long-term agreement will reassure the supplier of the manufacturer's commitment and will allow the supplier to quote a volume price discount.

2. Give the supplier a monthly forecast for a rolling window of six months. The supplier will use this forecast only for material planning purposes. The forecast can be changed within the agreed lead time specifications.

3. Give the supplier a monthly firm release for the next month of production.

4. Establish with the supplier the rate at which products are to be delivered to the factory (e.g., hourly, daily, or weekly).

5. Establish an agreement with the supplier on the policy for changing delivery rates. This policy should be very clear and must include both increases and decreases. The supplier's lead time will be the critical factor influencing the possible rate of change. For example, a typical schedule could be: plus/minus 10 percent the first month, plus/minus 25 percent the second month, and plus/minus 50 percent the third month.

Figure 6.1 shows an MRP system for procuring parts from a supplier. The output of the MRP is a series of order points for parts grouped in time periods called buckets. The materials planner determines the build schedules and the bucket time windows. The planner usually operates in weekly buckets. The MRP then calculates the part requirements based on the production master schedule's demand and the parts at hand in the factory. The order

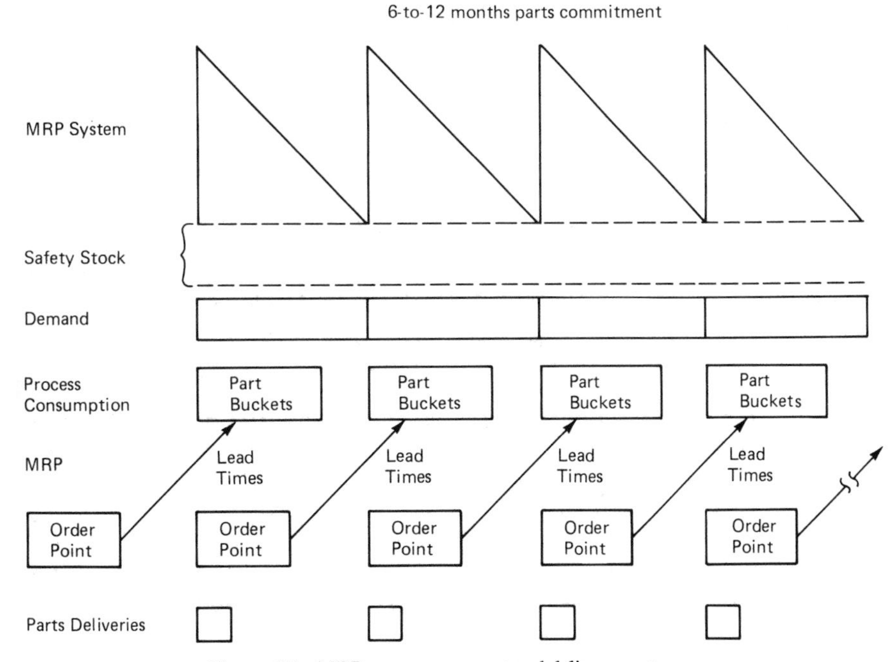

Figure 6.1 MRP part procurement and delivery system.

point for a part includes the supplier's lead time and the lead time of the process in the factory.

One of the problems with an MRP system is the long supplier lead times, which add inertia to the system. Also the concept of time buckets allows very little fine-tuning of any rescheduling. The system requires substantial paperwork and a lot of attention on the part of the planners and buyers. Finally, an MRP system operates in a push mode, causing excess material to accumulate in the factory.

Figure 6.2 shows a parts procurement system under Just-In-Time. This system is lean, effective, and quick. The buyer makes an eighteen- to twenty-four-month commitment to the supplier, and the planner provides six-month forecasts. Then the planner provides monthly releases based on the factory's usage rate. Figure 6.2 also shows the rescheduling policy mentioned above. The system operates in a pull mode, and it could use the MRP's output to forecast part requirements based on the master schedule demand.

Supplier's Lead Times

One of the biggest obstacles to implementing a Just-In-Time supplier program is the reduction of the suppliers' lead times. A solution to this problem must be devised before the program is implemented.

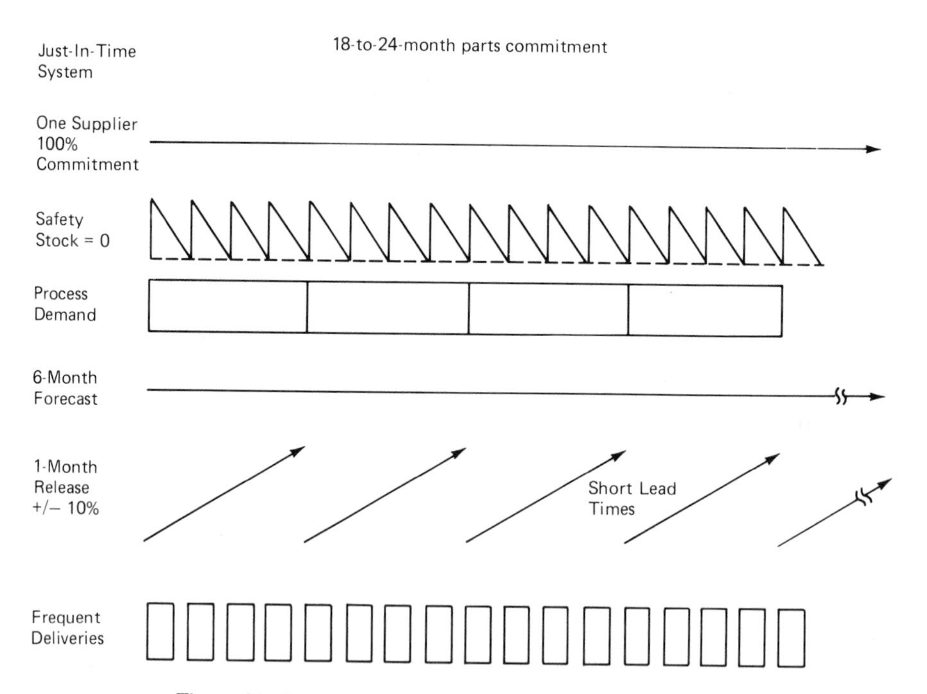

Figure 6.2 Just-In-Time supplier procurement and delivery system.

The first objection that a supplier is going to raise against reducing lead times is that there are long lead times for some of the parts it buys to build the products it sends to the manufacturer. The shorter lead times have a ripple effect through many layers of suppliers, complicating the process for the manufacturer's supplier.

One way to solve this problem is to ask the direct supplier to work with its suppliers to shorten the lead times of the parts affecting its lead time. A smart supplier will solve the problem this way and thus benefit from a Just-In-Time system. Instituting this solution is normally a slow process and requires a lot of coordination and technical support.

Another quick way to solve the problem is to ask the supplier to make an analysis of its long-lead parts, stocking enough of them in advance to be able to meet the manufacturer's Just-In-Time requirements. For example, let's assume that a supplier is shipping 2,000 units per month. Also, there are five parts in the supplier's product with lead times of 120 days. The supplier could put in place for each a buffer of 4,000 units in order to respond to a sudden 50 percent increase in demand.

Having a buffer inventory of critical parts should only be a temporary solution until the supplier's lead times are reduced. The investment the supplier is making in a buffer inventory to shorten its lead times creates a good incentive to go back and negotiate shorter lead times with its own suppliers. The idea is that in the end everybody will benefit from cost reductions from the improved system.

Teaming up with Suppliers

One of the main goals of a Just-In-Time supplier program is to reduce the number of suppliers that contribute parts to the product. This entails a manufacturer's developing long-term partnerships with a few single-source suppliers. A side benefit is the reduction of the purchasing and technical support overhead required to interface with suppliers.

In order to achieve this goal, a manufacturer must select suppliers not only for their prices but also for their willingness to work with the manufacturer in a Just-In-Time program. For example, a supplier must be able to accept the manufacturer's value analysis of its product in order to allow it to stay competitive. The value analysis is a good tool to motivate a supplier to lower its costs, for the manufacturer is making a long-term commitment that requires the supplier to be sensitive to cost-saving issues.

Another factor to be considered in selecting suppliers is their physical distance from the factory. Ideally, suppliers should be close to the factory because of the frequent deliveries and the need for communication and feedback concerning quality. All things being equal, it is better to select a nearby supplier rather than one located farther away.

To increase efficiency, it also makes sense to use clusters of suppliers. If the manufacturer selects suppliers close to each other, it can combine their deliveries using common carriers. This will save overhead in dealing with carriers. It will also produce savings in rates because of the increased volume of business with each of a smaller number of carriers.

6.2 QUALITY AND SUPPLIER PERFORMANCE

As noted many times, Just-In-Time is a crusade against waste. Inspecting a part doesn't add any value to it, for the inspection only confirms what should have been known from the beginning: The supplier shipped a part that conforms to specifications.

One of the major steps required by a Just-In-Time supplier program is the elimination of receiving inspection. When a part arrives to the receiving dock, it is routed directly to the manufacturing floor without any waste of time and labor. Routing dock to WIP obviously necessitates that high-quality parts are consistently received.

Chapter 8 addresses the concept of a TQC program and provides guidelines for implementation. It is strongly recommended that a TQC program be established with a supplier before beginning Just-In-Time deliveries. Also, TQC can be applied not only to suppliers but also to the factory. Having no buffer inventories entails the factory has no spare parts to cover for bad ones in the manufacturing process. A first-rate quality control system is a must if the manufacturer is to survive instituting such an operating mode.

Before talking to a supplier about Just-In-Time deliveries and TQC practices, the manufacturer should have a detailed quality record of the supplier. This record should include the last three to six months. Nothing is more effective than presenting the supplier with a chart showing the quality track record of the product received. Also, the manufacturer must be ready to commit its design and manufacturing engineers to helping the supplier improve quality. Once the quality meets the standards, the manufacturer can start reducing its receiving inspection and eventually cut it out entirely.

6.3 JUST-IN-TIME PURCHASING BENEFITS

This section lists some of the key benefits resulting from a Just-In-Time supplier program. In practice, a manufacturer will not achieve these benefits overnight. Hard work and lots of communication are needed before it will see results.

One of the things the manufacturer has to face in implementing this program is that some suppliers will have to be dropped in order to reach the goals. Just-In-Time calls for a reduction in the supplier base, and the selection of the good suppliers will take considerable time and effort. The manufacturer should not consider price as the only determining factor. Quality and a willingness to deliver small lots on time must weigh heavily in the selection of suppliers.

Cost of Parts

There are many costs associated with the use of parts in a manufacturing organization. The most obvious one is inventory carrying costs. A Just-In-Time supplier program—in which small lots are delivered frequently and on time—surely contributes to reducing inventories. Also, fine-tuning deliveries of parts to actual consumption rates minimizes the excess inventory in cases where something goes wrong with the process.

Another saving to consider is the decreasing cost of parts resulting from long-term commitments to suppliers. Volume purchases, suppliers' learning curves, and productivity increases will certainly add to savings. Conversely, in a traditional second-source approach, a manufacturer buys a part from several suppliers in order to have a backup source. This approach leads to lower-volume purchases and a corresponding increase in overhead (for the manufacturer has to deal with more than one supplier).

The concept of a single source is very important and should be discussed with the supplier during the Just-In-Time supplier program. There must be a clear understanding that the manufacturer expects cost reductions as volume increases.

Finally, there are cost savings related to rework and scrap. Since the supplier delivers parts in small quantities and there are no buffer inventories, the manufacturer can detect problems with parts early in the process. This avoids expensive rework costs, but it also requires a system for quick feedback to the supplier. The manufacturer has no spare parts to cover for the defective ones and it must urge the supplier to correct the problem immediately. This is one of the benefits—and risks—of the Just-In-Time system. The manufacturer puts itself against the wall and has to solve problems as they occur in order to keep the line moving.

Quality

Quality improvement is one of the most important benefits of a Just-In-Time supplier program. Parts purchased from Just-In-Time suppliers are of higher quality than those from regular suppliers. These parts in turn will produce higher-quality products to be shipped out of the factory. Frequent deliveries will contribute to the early detection of problems. Small-lot production runs and short setup times will also help the supplier to correct the problems quickly. Finally, the elimination of receiving inspection and the focus on process control switches the responsibility for quality to the makers of the parts.

Productivity Increases in the Purchasing Department

Having a small supplier base will help to reduce competitive bid activities in the purchasing department. Buyers will negotiate extended contracts with suppliers looking for long-term relationships. Frequent, on-time deliveries reduce the time a buyer has to spend expediting. Also, if the manufacturer selects suppliers close to its factory, it will substantially reduce traveling and communication expenses. For example, the overhead due to dealing with an overseas supplier will be much larger than for a local supplier. Using an overseas supplier will require long distance phone calls and large travel expenses. Also, there will be problems arising from the differences in culture and operating modes, the language barrier, and the length of the pipeline required to execute Just-In-Time deliveries. The next chapter discusses the concept of the in-plant store, which can further increase the productivity of the purchasing department.

The productivity of buyers will also be increased as a result of the numerous measures available to track a supplier's performance. Once the supplier is in the Just-In-Time program, it will ship products in standard containers, frequently, and in smaller lots; there will be an increase in quality and a decrease in paperwork. In consequence, the buyer will be able to detect and measure any potential problem with the supplier quickly and precisely. For example, assume that a supplier is shipping 200 parts per day but misses a shipment. The buyer will immediately know there is a problem with the supplier. This detection occurs faster than if the supplier missed a monthly shipment. In the latter case, the buyer would have to wait a full month before finding out the supplier is in trouble.

Design Improvements

A Just-In-Time supplier program requires close technical cooperation with suppliers. This cooperation involves not only manufacturing support engineers but also design engineers. Just-In-Time handles design changes and cost-saving improvements more efficiently, because small lots and frequent deliveries speed any feedback concerning engineering problems to the suppliers. Also, manufacturing can implement engineering-change rework faster, because there is less inventory sitting around. Finally, product specs must be clearer and simpler in a Just-In-Time system, for the supplier has to conform to the specs consistently in order for receiving inspection to be cut out.

Overall Productivity Improvement

In general, there is an overall improvement in productivity in the materials department once a manufacturer implements a Just-In-Time supplier program. It is true that the program's transitional phase will require more work and attention to detail. But in its steady state, Just-In-Time will clearly generate savings with respect to every function of the department.

Among the most productive changes are the switch to small lots of parts and the elimination of receiving inspections. An increase in the quality of the parts reduces the defect rates in the factory, with a corresponding increase in yields and a reduction in rework and scrap. Linearity in delivering parts and in the factory produces even production rates and less need to work overtime to catch up with last-minute slippages in the schedule. Finally, the improved quality of the product will translate into a higher degree of customer satisfaction and a reduction in field service activity.

Just-In-Time versus Traditional Batch-oriented Procurement

Following is a list of the fundamental characteristics of a traditional procurement system and a Just-in-Time supplier program. The list provides a quick glimpse at the operating differences between the systems.

TRADITIONAL PROCUREMENT	JUST-IN-TIME PROCUREMENT
Weekly/monthly deliveries	Hourly/daily/weekly deliveries
Buffer inventories	No buffer inventories
Acceptable quality rejects	Total quality control and zero defects
Long lead times	Short lead times
Early/late deliveries	On-time deliveries
Large lots deliveries	Small lot deliveries
Receiving inspection	No receiving inspection
Dock-to-stock transfers	Dock-to-WIP transfers
Extensive expediting	Relatively little expediting
Inconsistent packing	Standard packing
Second-source suppliers	Single-source suppliers
Large supplier base	Small supplier base
No physical distance restrictions	Nearby suppliers
Short-term hard releases with purchase orders	Long-term commitments and short-term forecasts and releases
Adversarial relationship with suppliers	Partnership relationship with suppliers
Random feedback on quality	Continuous feedback on quality
Infrequent visits to suppliers	Frequent visits to suppliers

6.4 HOW TO START A JUST-IN-TIME SUPPLIER PROGRAM

This section reviews the necessary steps required to start a Just-In-Time supplier program. First, let us define when a supplier is in the program:

- The supplier commits itself to deliver the product according to a repetitive, small-lot schedule. Depending on the volume, deliveries might be hourly, daily, or weekly.
- The supplier might have to commit itself to maintaining a buffer inventory of long-lead parts in order to reduce its lead times so as to be in accord with the program's rescheduling policies. The size of the buffer inventory will depend on the rolling six-month forecasts.
- The manufacturer will make a long-term commitment to the supplier in order to motivate the supplier's acceptance of the plan. Depending on the life cycle of the

manufacturer's product, a commitment of between eighteen and twenty-four months would be ideal.

- A TQC program with the supplier must be in effect. The objectives of the program are twofold: (1) to bring the supplier's quality level up to the point where it consistently delivers high-quality parts and (2) to end the need for receiving inspections of the supplier's shipments. After inspections are stopped, the receiving clerk will route the parts directly to the process, bypassing the stockroom. A quality and problem feedback system must be in place.
- The supplier receives periodic quality and reliability reports on its product's performance in our process and in the field.
- Supplier's source inspection is only acceptable as a transitional phase of the Just-In-Time program.
- We will make it very prestigious for a supplier to belong to the Just-In-Time program. This requires some type of supplier certification and award recognition program.

Implementation Steps

Below is a list containing the nine steps required to implement a Just-In-Time supplier program. Some of these steps are discussed in the sections that follow. Also, Chapters 5, 7, and 8 deal with lead time reduction, frequent deliveries, and quality improvement programs. On the whole, every company needs to tailor a Just-In-Time supplier program to its own product and operational arrangements. The objective of presenting this list is merely to offer a general outline for implementation.

JIT-TQC SUPPLIER PROGRAM MILESTONES

1. Select suppliers for program.
2. Institute JIT-TQC supplier education program.
3. Work out JIT contract.
4. Implement TQC programs:
   - quality improvement program.
   - process improvement program.
   - lead time reduction program.
5. Certify JIT-TQC suppliers.
6. Hold a Suppliers Day.
7. Eliminate receiving inspections.
8. Begin frequent deliveries in small lots under a JIT pull system.
9. Evaluate JIT program performance and make necessary changes.

Education and management commitment. There are several steps that are preliminary to the implementation of a JIT-TQC supplier program. The first is to educate the team responsible for implementation. It is very important that everybody on the team have a clear understanding of the supplier program principles and goals. Every manager, including those in top management, must be committed to supporting the effort.

The simplest way to train employees is to send them to one of the many seminars available throughout the country. If the group is big enough, it is better to establish an in-house education program or to train several employees who in turn will train the rest. Top management should participate in some of the seminars so that they understand the program's benefits and feel comfortable supporting it. Finally, it helps to put together a Just-In-Time library, where books, articles, and visual training aids are available to those working on the program.

Description of the program and its goals. The next preliminary step is to put in writing a description of the Just-In-Time supplier program. This document should clearly state the program goals. For example, it should describe the scope of the program, the desired number of suppliers, and the main milestones in the process of getting suppliers actively participating. The program should not be too ambitious or the goals unreachable. It is important to start with only a few suppliers in order to allow learning by experience and to provide time to make changes as problems arise.

The program milestones will help in monitoring progress and in concentrating efforts in problem areas. Weekly meetings between management and those involved in implementing the system are recommended in order to review progress.

The plan should include milestones in three important areas: (1) supplier selection and incorporation into the program; (2) education of suppliers about the TQC program; and (3) the changing of factory processes and paperwork in order to operate in a Just-In-Time pull mode.

List of supplier candidates for the program. The most effective way to select suppliers for the Just-In-Time program is to rank them by dollar volume and importance to the manufacturer's production needs. Usually inventory class A and B suppliers are the best candidates. These suppliers contribute the highest-dollar or highest-volume parts to the production line. The list of candidates should contain about twice as many as will be chosen, leaving a margin for dropouts. Also, a complete list of all suppliers grouped by commodity should be constructed. This helps in eliminating multiple source suppliers. Having a single source for each part is an important goal and it should help in deciding which suppliers should be in the program. A strong deciding factor is whether a supplier is willing to cooperate with the manufacturer.

The next step is to make a table containing the supplier candidates. For each supplier, the table should include the volume of parts the company purchases from the supplier, the lead times, the quality levels, the delivery track record, the current delivery schedules, and any pending problems with the supplier's performance. The management team then prioritizes the suppliers by order of importance to the program. Once the table is complete, the suppliers are contacted in the order in which they appear on the prioritized list.

Just-In-Time supplier contract and the certification program. An important step in the implementation of a Just-In-Time supplier program is the signing of a Just-In-Time agreement with each supplier. The agreement should outline the agreed operational arrangements for receiving Just-In-Time shipments. Appendix A shows a sample agreement used at Metaphor Computer Systems. This agreement was slightly modified for every supplier in the program to accommodate previous OEM contracts and specific operating modes. The sample agreement is a good starting point, but there are numerous possible variations.

Any agreement should include long-term commitments, timely rolling forecasts, monthly releases, delivery rate schedules, lead-time commitments, quality levels, price and price negotiation time frames, and single-source commitments. Every supplier signed to the program must commit itself to specific dates by which to start delivering products in a Just-In-Time fashion.

In addition to the agreement, a Just-In-Time certification program for the suppliers needs to be instituted. This program will grant to a certified Just-In-Time supplier the long-term, single-source status described in the agreement. The manufacturer should make it very prestigious for a supplier to belong to this program, issuing certificates and displaying copies of them in its lobby. Also, the manufacturer should make clear that a supplier could get disqualified if quality, delivery, or performance fall below certain levels. Once a supplier is disqualified, the manufacturer needs to establish with the supplier a remedial program to correct the problems. If the program fails, the manufacturer should consider getting a new supplier.

Supplier certification and Suppliers Day. Once the manufacturer has selected the group of suppliers to enter the program, it should notify them in writing of its intentions and invite them to a Suppliers Day. The invitation letter should list only the general goals of the program, leaving the details for the Suppliers Day meeting. The letter should also ask the supplier to assign one of its employees to be the coordinator of the Just-In-Time supplier program.

A typical Suppliers Day agenda is shown below. In general, the purpose of a Suppliers Day is to explain to the suppliers the scope of the program and the steps required to implement it. Also, the manufacturer should stress its commitment to the program.

SUPPLIERS DAY AGENDA

- Registration
- Welcome (high level management, vice president)
- Introductions and agenda
- The Just-In-Time system and the manufacturing challenge
- Break

- Just-In-Time purchasing
- Total quality control
- Lunch
- Just-In-Time supplier program phases and implementation; certification program
- Company's market outlook and one- to three-year prospects (vice president or president)
- Break
- Question and answer session
- Cocktails and hors d'oeuvres

It is important that top management participate to demonstrate the manufacturer's commitment. Suppliers must leave the meeting with a clear understanding of the manufacturer's expectations. The manufacturer should make sure that suppliers understand the challenge and the overall benefits they will gain. Once the suppliers have learned the Just-In-Time principles, they can apply them to their own suppliers. Below is a list of expectations the company and the suppliers should have regarding the Just-In-Time supplier program.

JUST-IN-TIME SUPPLIER PROGRAM EXPECTATIONS

1. What suppliers should expect from the manufacturer:

   - long-term business partnership
   - fair dealing
   - fair profits
   - single sourcing
   - customer satisfaction
   - fair return on investment
   - enough time to plan ahead
   - accurate and timely forecasts
   - parts specifications called out correctly
   - parts designed to match the manufacturer's process
   - smooth order releases
   - minimum order changes
   - prompt payment of invoices
   - quick feedback on quality problems
   - good technical support to solve quality or process problems
   - frequent communication

2. What the manufacturer should expect from the suppliers:

- high-quality parts that meet requirements consistently
- on-time shipments
- hourly/daily/weekly deliveries
- short lead times
- full-service supplying
- quick feedback on problems and repairs
- fair dealing
- long-term business partnership
- excellent process control to ensure quality levels
- quality of parts sufficient to dispense with receiving inspections and buffer inventories
- cost reductions and learning curve savings passed on to the manufacturer
- parts packed and shipped in convenient quantities and packaging
- frequent communication
- open channels for technical support in case of quality problems

6.5 SUMMARY

A Just-In-Time supplier program is difficult to implement. Normally when a manufacturer implements Just-In-Time internally, it has the control necessary for success. In a Just-In-Time supplier program, the manufacturer is dealing with other companies that have different operating environments. These companies most likely will resist changes, even if the changes would improve their systems.

A manufacturer should consider carefully before implementing an aggressive supplier program, for it might require more resources than are available. The manufacturer should determine all the resources that would be needed, particularly at the beginning, when technical staff often must take time to sort out quality problems and work with the suppliers to correct them.

Every supplier has a unique operating environment and manufacturer officials should keep an open mind in setting up a Just-In-Time supplier program. The manufacturer must maintain a certain degree of flexibility in dealing with suppliers. This doesn't mean, however, that the manufacturer should compromise its Just-In-Time goals to accommodate a supplier's inefficiencies. It must stand firm regarding the basic Just-In-Time principles and switch to another supplier if it reaches a stone wall. The process of replacing a supplier might take time, but it will prove beneficial in the long run.

The final piece of advice is to define the supplier program well in advance and then assemble a good team to carry it out. The team members must clearly understand the program's goals and principles. Then they should work consistently with suppliers until together they polish the system by applying their accumulated experience.

REFERENCES

BELL, ROBERT R. "Managing Change in Manufacturing." *Production and Inventory Management* (American Production and Inventory Control Society) 28, no. 1 (1987):106–114.

CONSTANZA, JOHN R., and DAVID R. WAGNER. "J-I-T Purchasing and Source Management." In *Just-In-Time Manufacturing Excellence*. Published by the authors, 1986.

FALLON, MICHAEL R. "Customer-Supplier Relations and JIT: What's Needed for Success." *Managing Automation,* January 1987, pp. 40–44.

HEARD, ED, and GEORGE PLOSSL. "Lead Times Revised." *Production and Inventory Management* (American Production and Inventory Control Society) 25, no. 3 (1984): pp. 32–47.

PLENERT, GERHARD. "Are Japanese Production Methods Applicable in The United States?" *Production and Inventory Management* (American Production and Inventory Control Society) 26, no. 2 (1985): pp. 121–129.

SCHONBERGER, RICHARD J. "Just-In-Time Purchasing." In *Japanese Manufacturing Techniques: Nine Hidden Lessons in Simplicity*. New York: The Free Press, 1982.

7 ★

Moving Vendors into the Factory: The In-Plant Store

Just-In-Time calls for suppliers to deliver frequent, small lots of high-quality parts based on the customer's consumption rates. This requires frequent communication and a close relationship between a manufacturer and its suppliers. One way to minimize the overhead associated with the additional activity is to reduce the number of suppliers and to select suppliers relatively close to the factory.

In-plant stores offer an ideal solution. In an in-plant store program, a supplier teams up with the customer in the planning, procuring, and stocking of parts at the customer site. The program calls for the supplier to open a dedicated store on the customer's premises to support its materials needs. Once in the same building, the daily operation becomes routine and efficient. The in-plant store brings the supplier to within walking distance of the customer and facilitates frequent deliveries directly to the production line.

The procurement of electronic parts from electronic distributors, for example, lends itself to being done through an in-plant store program. In general, computer systems companies purchase from distributors hundreds of different electronic parts to assemble the electronic modules that compose their products. Purchasing these parts is a complex process, because it involves a wide variety of semiconductor companies, many of which will not deal directly with relatively small customers. Electronic distributors provide the advantage of allowing parts from many semiconductor companies to be purchased from a single supplier. They also provide stocking services and the economy of scale for the pricing of parts, for they buy parts from the semiconductor suppliers in volumes equivalent to the combined orders of all their customers.

Hamilton/Avnet, one of the biggest electronic distributors in the United States, pioneered the idea of the in-plant store. Metaphor Computer Systems has had an in-plant store run by Hamilton/Avnet for several years (Figure 7.1). The arrangement calls for Hamilton/Avnet to stock electronics parts used in Metaphor's products and to manage the planning, procurement, and kitting of them. Hamilton/Avnet also agreed to maintain a buffer inventory and to staff the in-plant store. The program is a true partnership between Hamilton/Avnet and Metaphor.

This chapter describes the principles of the in-plant store as both companies have modeled it. In practice, the concept can be applied to other situations as long as the basic operating mode of the store is followed. The idea is to develop a partnership between the supplier and the customer. The supplier gains a customer who buys as much product as possible from the store. The customer gains a reliable supplier that participates in materials management within the company.

Figure 7.1 Hamilton/Avnet in-plant store at Metaphor Computer Systems.

7.1 IN-PLANT STORE OPERATING RULES

It is critical that both partners in an in-plant store program gain from the relationship without restricting the competitiveness of their businesses. For example, the customer would like to pay fair market prices for the parts in spite of the additional services that the in-plant store offers. At the same time, the supplier would like to recover the additional overhead expenses required to stock and manage the store. The in-plant store operating arrangement should allow both partners to reach their goals. In other words, they should have a win-win relationship.

The in-plant store operating rules are listed below. In general, they could be used as a starting point for similar arrangements involving other types of businesses.

IN-PLANT STORE BASIC RULES

- The supplier will provide a buffer inventory based on the customer's forecast.
- The supplier will staff the in-plant store and kit the parts before delivering them to the customer.
- The supplier will receive its own parts when they arrive at the customer's receiving dock.
- The supplier will base its prices on the fair market value of the parts.
- The supplier will install communication equipment so that the store manager can place orders at the supplier's main business location.
- The customer will commit to a certain annual volume of business with the supplier.
- The customer will give the supplier the right of first refusal on new parts released to manufacturing.
- The customer will provide a secure room for stock free of charge.
- The customer will provide a six-month rolling forecast updated every month.
- The customer will provide free computer services to allow the supplier to run MRP and to do kit pulling.
- The customer will provide training and proper documentation on its product line and the use of the customer's computer facilities.
- A sale is transacted when the supplier delivers the parts to the customer.

This list can be the starting point in developing the first draft of an in-plant store contract. The supplier and the customer must write a contract before opening the store. The contract will become the operating reference for managing the store.

It is recommended that the initial length of the contract be two years, renewable in one-year increments. Also, the contract should be flexible enough to allow addendums as new needs arise during that time. It is also recommended that the contract include a list of the parts supplied by the store and the agreed prices. Later on, this list could be updated and parts added or deleted as the products change.

7.2 MATERIALS PLANNING AND PURCHASING

Essential to the success of the in-plant store is the planning for materials needed to support the customer's production line. A good in-plant store will assume the responsibility for materials planning and purchasing. The store manager must be capable of planning materials in the same fashion that a materials planner would in the customer's materials department. This is an area where the in-plant store could have problems. Suppliers normally don't have the experienced staff or the computer tools needed to do a good job in planning material requirements. The customer must help the supplier in this area to ensure the success of the store. For example, the supplier must have software capabilities for running MRP and for tracking purchase orders and inventory. Also, there must be software tools to produce kit lists to pull the materials from their stock locations. All of this requires special attention and a joint effort to avoid having the store fail.

Materials Forecast

In the in-plant store contract, the customer agrees to provide the supplier with a rolling forecast of parts for a particular time window. A typical forecast should cover a window of six months, with a front end defined as nonchangeable. The customer should have a thirty-day firm release in the front. This will provide the store manager with some schedule of the parts he or she needs to kit and deliver.

The store manager will use the forecast as a materials planning tool. But the forecast should not include every part the store is supplying. The process can be simplified by grouping the discrete parts into kits that the store will normally deliver. This step is very important as a way of streamlining the paperwork of the store manager and of the planner interfacing with the store. For example, assume the customer has several printed circuit boards with electronic components. The forecast will only list the top assembly numbers of the bill of materials. The bill of materials in turn will list all the components used in the printed circuit boards.

Before opening the store, the customer planner and the store manager should agree on the data base format for the parts and the forecast. They should concentrate their efforts on simplifying the information required to operate the store. The focus should be on a simple, consistent forecast that would give the store as much visibility as possible.

MRP and Purchasing

The in-plant store simplifies material requirement planning on the customer's side. The process will be reduced to the same extent that the number of parts that the store handles is reduced. The MRP will not be required to explode the store parts and need only reach the top assembly numbers used to provide a forecast of the parts to the store.

During computing, the MRP has to ignore the store's component lead times and consider only the time it takes the store to service a kit request. Then the in-plant store has to consider the lead times of the individual parts when it is running its own MRP. For example, if the customer is buying five different kits from the in-plant store and the store's turnaround time is five days, then the lead time in the customer's MRP should be five days. Let's assume that these five assemblies consist of four hundred components. The in-plant store will have to enter the lead times of each one of those components on the store data base so that the MRP can calculate their order points. This process places the burden of planning parts on the in-plant store.

One of the main goals of the in-plant store is to simplify paperwork and reduce the related overhead. The workload of issuing and tracking purchasing orders is minimized with the in-plant store. Referring to the example above, there would be no need to write a purchase order for every one of the four hundred components composing the bill of materials. A simple contract listing the agreed price per part and the parts' annual volumes is sufficient. Then when the store delivers a kit to the customer, the release paperwork can function as the invoice for that particular set of parts. The store manager will send the invoice to the accounting department for payment without the need for a complex purchase order system. This reduces the need to write purchase orders and simplifies the changes associated with adjustments in production schedules. Also, tracking parts becomes very simple, for the customer is taking possession of the parts at the time the store delivers them. This arrangement has worked very well for Metaphor and Hamilton/Avnet (their in-plant store contract specifies payment terms and discount rates). The benefits of a simplified purchasing arrangement include fewer purchase orders to write, fewer buyers to support the system, and the elimination of overhead and time waste.

Buffer Inventories

Just-In-Time calls for zero buffer inventories. This is a goal that should always be kept in mind by those involved in managing the system. In some cases, the variety of parts and their long lead times make it very difficult to support a manufacturing process without a buffer inventory. The ideal solution is to have the supplier carry such an inventory (e.g., in an in-plant store).

The size of the buffer inventory is one of the critical issues to negotiate in devising an in-plant store agreement. Usually a manufacturer has a general idea of the buffer needed to support the process. A simple way to measure buffer size is in weeks of parts at hand based on the current production schedules. At the low end, the manufacturer can ask the in-plant store to carry two weeks of buffer inventory at whatever consumption rate is effective at the time. At the high end, the manufacturer can request twelve weeks. A good compromise is a six-week buffer. One important point to discuss at contract time is the commitment that the manufacturer has to this buffer inventory. In most cases there is no need to have a firm commitment to take the buffer inventory if the parts are off-the-shelf returnable by the supplier. But for cases in which the supplier is bringing custom parts made to the manufacturer's specifications, the manufacturer must commit to take these parts no matter what happens to its demand. In this case, the material planners must exercise caution and adequately forecast the need for the parts.

Having a buffer inventory provided by the supplier is one of the main benefits of the in-plant store. It is very important that the manufacturer negotiate its size before finalizing the store contract. Any material that the supplier agrees to store for the manufacturer will mean a reduction in its carrying inventory. It also means the store will probably have no shortages and be able to deliver parts on time.

7.3 IN-PLANT STORE OPERATING ARRANGEMENTS

Before opening the in-plant store, the store operating agreement must be set down in writing. This agreement will guide the operation of the store and will help to resolve any conflict that might arise. The goal is to solve conflicts at the lowest working level and to involve management only on rare occasions.

Product Pricing

Product pricing is a very important issue during the negotiation of the contract for the in-plant store. The supplier will be sensitive to the need to recover the overhead invested in the store. This overhead relates to the investment in the buffer inventory and in the staffing of the store. On the other hand, the manufacturer will not want to pay above market prices for parts bought from the store.

The conflict can be solved by the parties agreeing to two conditions. First, the manufacturer should pay the fair market price of the part based on quotes from other suppliers. These quotes usually give a good idea of the price of the part and can serve as a reference point for the store's pricing. It is part of this first condition that the manufacturer should have the right to buy the part from another source if the in-plant store's price is higher. This will give the supplier an incentive to match the open market price of the part.

The second condition is favorable to the in-plant store. The manufacturer should make a purchase commitment for a certain annual volume. This purchase volume should be based on the supplier's model of the store and should be calculated so as to make the enterprise profitable. Also, the manufacturer should grant the store the right of first refusal on any new part released to manufacturing. Under this clause, the store will get the business on any new part if the store matches its fair market price. To keep the system honest, it is very important that the manufacturer make a conscious effort to buy as many parts through the store as possible.

Once the price for every part is listed on the contract, the manufacturer's buyers must monitor those prices on the market and watch for price changes. Every quarter, the buyers should check the most volume- and price-sensitive items. The manufacturer will then pass this information to the supplier and request a new quote on the parts. In most cases, the in-plant store will match the new price.

If the store supplies custom parts from only one original supplier, the manufacturer should first negotiate directly with the original supplier. It can then negotiate with the store the mark-up percentage to cover the handling and stocking costs for the parts. For example, assume that the manufacturer negotiates a $200 part price from a supplier in quantities of 5000. It could agree to let the store handle the parts for a 8 percent mark-up. The size of the buffer inventory will be a major contributing factor regarding the mark-up percentage.

Dealing with Shortages

Part shortages can easily cause friction between the in-plant store and the operating staff in the materials department. The in-plant store contract should specify the lead time for the store to deliver parts. For example, the contract could specify a five-day lead time to pull and deliver a kit after it has been requested.

In a normal operating environment, there will be times when a part shortage prevents the store from delivering a kit within the guidelines. In that case, the manufacturer should not take the kit until the store has filled all the parts.

The most effective way to monitor store shortages is to invite the store manager to report the status of the released kits in the materials department shortage meetings. By participating in the meetings, the manager will become familiar with the department staff and feel part of the team. The meetings will also provide advance notice of problems with parts and of situations in which the store won't be able to deliver on time.

A backup solution to the shortage problem is for the manufacturer to preserve the right to fill any shortages from other sources. Only if the manufacturer decided to exercise that option would it take a short kit from the store.

In practice, the manufacturer should not exercise that option often, because its purchasing department will begin to resent filling shortages for the store. Also, filling such shortages might make the store manager feel less responsible for procuring the parts. Shortages in the store kits are a good index of the store's proficiency. Metaphor's experience with Hamilton/Avnet has been very good. The in-plant store is on line with Hamilton/Avnet's distributor network, which makes the store manager very efficient at resolving part shortages.

Purchase Orders and Accounting

The in-plant store offers a shorter path to the source of the part than in a normal procurement cycle. This advantage translates into a reduction in the time it takes to procure parts. Also, there is a reduction in the number of items that the manufacturer's planners and buyers will handle.

For clarity, let's review the differences between buying a part through the in-plant store and buying it through the purchasing department. Normally, the need to buy a part starts with the production master schedule driving the MRP system. The MRP checks the stock at hand and the lead time of the part. Then it tells the planners when to order the part and what quantities are required. The MRP specifies these orders points in a time-phased schedule. The planner fills out a purchase requisition, which describes the order quantities and delivery dates. Then the planner passes the purchase requisition to the purchasing department. The buyer contacts a salesperson in the company that sells the part and requests a quote and delivery dates. The buyer can also contact other suppliers to get competitive bids. The salesperson in each company contacted passes the request to the internal order department and asks for a quote and a delivery commitment.

The buyer generally places the order with the lowest bidder and issues a purchase order to the supplier. A purchasing clerk will enter the purchase order into the computer system so that when the part arrives it has matching paperwork.

When the part arrives at the receiving department, the receiving clerk matches the part to the purchase order entered on the computer. Then a receiving inspection will handle any discrepancies in quantity or quality. The buyer will get involved in solving any problems with the order. The receiving clerk also sends a copy of the receiver receipt to accounting to wait for the invoice for the part. As can be seen, this process takes a substantial amount of time and the involvement of a lot of people, and there is of course the corresponding overhead.

In an in-plant store system, the planner only worries about releasing the top bill of material that calls for the parts the store handles. Then the planner supplies the store manager with a rolling forecast based on the MRP demand, releasing kits daily or weekly depending on the need for the parts in the production line. The store manager plans and orders the parts without involving the purchasing department. Also, the store manager will receive the parts in the back door and takes responsibility for checking them against the order he or she had placed. The store provides great savings in paperwork and labor.

Another simplification in the procurement process is the elimination of purchase orders. The store contract will list the parts on the agreement together with their prices. When the store delivers a kit, the dollar value of the kit becomes the invoice that goes directly to the accounting department for payment. This eliminates the need to keep on the computer a detailed purchase order record for every part supplied by the store. The procedure is simple and effective. Accounting will pay the invoice using the guidelines specified on the contract (e.g., net thirty days or whatever discount agreement is in effect). The manufacturer must negotiate the store payment arrangement in advance and write the terms into the contract.

Staffing the In-Plant Store

The selection of the store's personnel is a key factor in determining whether it is successful. The store manager must have some background in materials planning and also be capable of placing orders with several suppliers. Depending on the size of the store, the manager might have to pull the parts into kits and to expedite the filling of shortages. The manager also has the responsibility of receiving parts and keeping inventory counts accurate. Finally, the manager represents the supplier in materials review meetings.

At Metaphor, the in-plant store was successfully integrated with the materials department by having the store manager participate in the daily materials meetings. In these meetings, the store manager reports on the status of the kits released to the store and on any potential problems that might result in shortages.

Finally, in selecting an in-plant store manager it is important to consider the likely reception by the materials department. Metaphor tried to avoid any problem by having the materials department take part in interviewing and selecting the store manager. This approach seems to have worked, for the materials organization has supported the manager they had helped to hire.

Kanban and the In-Plant Store

A Kanban system is ideal for regulating requests for material from the in-plant store. Metaphor developed a Kanban system between the in-plant store and the factory. When the process consumes a set of parts, the planner drops a Kanban into the in-plant store mailbox requesting a new kit. This system works in a pull fashion and in accordance with Just-In-Time guidelines. The small lot principle of the Kanban system also helped the store to increase the count accuracy of their kits, which ended the need for auditing at the time they are accepted.

The Kanban system also allowed the in-plant store to pull ahead one Kanban, so when the planner dropped a new Kanban in the mailbox, the store was ready with the parts. This simple procedure shortened the Kanban system's lead time from five days to less than one hour.

Bill of Materials Updates and Data Base

The need for proper levels of documentation is more critical with the in-plant store than with a regular supplier. In conventional purchasing, when the buyer purchases a part, the part's spec accompanies the purchase order without reference to the next assembly level that will use the part. In the in-plant store, the next level's bill of materials is needed in order to allow the store to plan, run the MRP, and pull the parts required by the Kanbans. To simplify the information, the manufacturer should structure the bill of materials into a single level. Also, it should not include in its bill of materials the parts called for on the store bill of materials. This arrangement will avoid duplication of requirements. Figure 7.2 shows the bill of materials arrangement for an in-plant store operation.

To keep the documentation current, the manufacturer must make sure the in-plant store receives all engineering changes issued against the bill of materials. Then the store manager has the responsibility of updating the information on his data base. One way to implement the bill of materials upgrades and to control their effectivity dates is to set up a separate data base for the in-plant store in the manufacturer's computer system. This data base must be independent from the manufacturer's but it must be capable of accepting electronic updates, including effectivity dates.

Engineering and the In-Plant Store

Engineering plays an important role in the success of the in-plant store. First, engineers must help structure product documentation so it identifies the parts the store supplies. The structure of the bill of materials should group the parts the store is supplying. Also, the store must

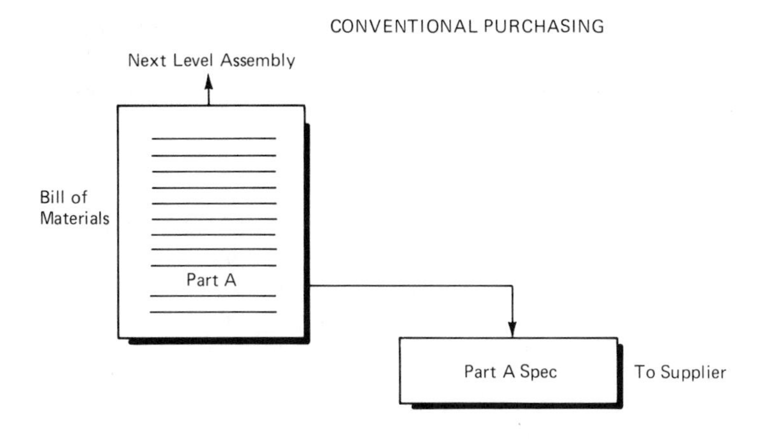

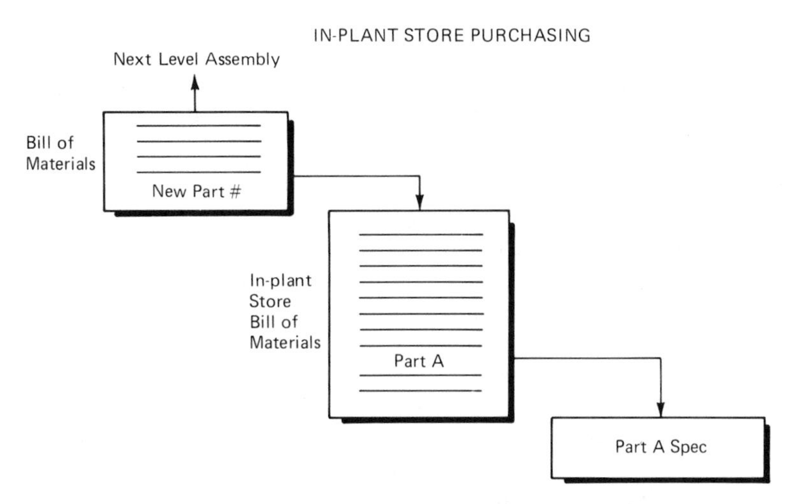

Figure 7.2 Bill of materials arrangement for an in-plant store.

have clear parts specifications so that there is no question about the type of part the bill of materials is referring to.

Normally, the planners in the materials department are responsible for seeing that the in-plant store gets regular engineering updates. The planners are also responsible for the in-plant store cut in dates for engineering changes and for coordinating new releases.

The second way in which engineering can contribute to the success of the in-plant store is by designing products that use parts the store can supply. The objective is to select those parts that will increase the purchase volume through the store. Every store part that the engineers include in their designs will reduce overhead in the materials department.

One side benefit for engineering is that it can use the in-plant store to procure sample parts and manufacturers' technical information. This service will save the engineers time and effort in contacting salespersons and writing purchase requests.

Quality and the In-Plant Store

The in-plant store is responsible for supplying parts that meet the quality standards of the Just-In-Time program. Also, the same TQC principles that are presented and discussed in the next chapter apply to the in-plant store supplier.

Receiving parts without inspection, vendor certification, and quality improvement programs must be coordinated with the quality control department of the company in which the in-plant store is located.

In the case of electronic distributors, a customer company must realize that they don't have quality control staff to address quality issues. This calls for close cooperation between the in-plant store, the electronic suppliers, and the quality department of the company where the in-plant store is operating. If the in-plant store delivers defective parts, the store will be responsible for the parts and must work with the original manufacturer to solve any problems. The quality department of the company where the store is located will serve as a technical advisor and will be the final judge as to whether the problem has been corrected.

Management Involvement

Top management support for the in-plant store is crucial for its success. One of the problems that a manufacturer has to overcome during the initial implementation is that the materials department will have reservations about committing to a single supplier. Department staff usually feel that it will be difficult to obtain service from other suppliers when the manufacturer needs parts to fill shortages or needs to find out market prices. The Just-In-Time principle that there should be only a few suppliers is worrisome to the materials staff. Their feelings of worry can get even worse if the supplier is moving in house.

One way to minimize the uneasiness is to involve the materials department management early in the negotiations concerning the in-plant store. It might also be possible to have the supplier take department managers and key employees on a tour of other in-plant stores. These visits would help allay the fears of the materials staff, for they could talk with staff of other materials departments about the impact of in-plant stores on relationships with external suppliers. This usually decreases the initial reservations about the store and gives managers and employees the opportunity to feel part of the implementation process.

After a year of operation, most of the concerns will prove to be unfounded, and the materials department staff will enthusiastically support the in-plant store concept. The savings in inventory, paperwork, and parts handling by far surpass any inconveniences in dealing with other suppliers.

7.4 IN-PLANT STORE ADVANTAGES

The in-plant store creates synergism between the supplier and the customer. By moving the supplier into the manufacturer's building, the factory can operate in a truly Just-In-Time mode. The store will reduce most of the time and material waste that occurs in a conventional procurement system. There is improvement in communication, overhead sharing, planning, buying, and material kitting. Also, the buffer inventory represents the supplier's financial commitment to supporting the manufacturer.

The simplification of paperwork and order processing increases productivity of workers and reduces the need for a large purchasing staff. The quick response to changes and problems eliminates the need for a large buffer inventory. The short distance to a supplier makes communication very effective.

The in-plant store is a natural for the distribution of electronic parts, since an electronics distributor supplies each customer with several hundred different parts from different suppliers. The nature of the distributor's business, which involves stocking most of those parts, will not make keeping an inventory at the customer's site a major burden. The Hamilton/Avnet store has been in operation at Metaphor Computer Systems for over four years and the results have been excellent.

7.5 SUMMARY

The in-plant store represents a new approach in the business of managing materials. The challenge for the future is to extend the concept to other types of activities, including manufacturing. For example, a supplier might have a small factory within the customer's factory and be responsible for a share of the manufacturing. In fact, in some cases a supplier's business might consist of a collection of small factories efficiently located on the premises of major customers. Such a supplier would essentially be a decentralized large business consisting of smaller, more efficient, and lower-overhead units that would build only what is needed, when it is needed, and with a minimum of waste. In other words, it would exemplify the Just-In-Time approach.

REFERENCES

ANSARI, A., and BATOUL MODARRESS. "The Potential Benefits of Just-In-Time Purchasing for U.S. Manufacturing." *Production and Inventory Management* (American Production and Inventory Control Society) 28, no. 2 (1987): pp. 30–35.

FALLON, MICHAEL R. "Customer-Supplier Relations and JIT: What's Needed for Success." *Managing Automation,* January 1987, pp. 40–44.

HERNANDEZ, ARNALDO. "Just-In-Time for a Start-up Company." Paper presented at the Society of Manufacturing Engineers. EMTAS 1986 Conference, Dearborn, Mi., March 1986.

MANOOCHEHRI, G. H. "Suppliers and the Just-In-Time Concept." *Journal of Purchasing and Materials Management,* Winter 1984, pp. 16–21.

SCHONBERGER, R. J., and P. J. GILBERT. "Just-In-Time Purchasing: A Challenge of U.S. Industry." *California Management Review* 26 (Fall 1983): pp. 54–68.

SCHONBERGER, R. J., and A. ANSARI. "Just-In-Time Purchasing Can Improve Quality." *Journal of Purchasing and Materials Management* 20 (Spring 1984): pp. 2 ff..

8 ★

Total Quality Control and Just-In-Time

In a Just-In-Time system, a manufacturer does not carry excess inventory to cover for parts that are defective. This forces the manufacturer to solve quality problems before it can continue building the product. To reduce the risk of frequent line stops, it must have a quality control program in place before reducing its inventory.

Under a total quality control (TQC) program, the manufacturer takes successive corrective actions regarding quality-related problems until the product conforms to the specifications 100 percent. TQC requires a commitment from every individual in the company to manage the quality control process so as to produce the best products possible. A TQC program is critical for the success of a Just-In-Time system, for no Just-In-Time system can function without high-quality parts and products.

A TQC program does not use inspection to ensure the quality of parts but rather it shifts the responsibility for quality to the makers of the parts. This puts an emphasis on prevention and on the implementation of a good process control system. It also requires a first-class information system to provide feedback on problems to those responsible for correcting them.

TQC is everybody's business, from the top of the company and those working on the production line. The responsibility for quality also extends to the suppliers, for the manufacturer will never produce a first-rate product unless the parts used to build the product are first-rate as well.

8.1 QUALITY AND JUST-IN-TIME: A TQC PROGRAM

The first step in developing a TQC program is to define the areas of responsibility for quality. This applies to the departments involved in the design, manufacturing, and support of the product. Once these departments have accepted their role in producing quality products, a TQC implementation plan containing specific guidelines can be devised. Below is a list of departments and their TQC responsibilities.

- The manufacturing department is responsible for the quality of the product shipped to the customer.
- The quality control department is responsible for setting quality standards and implementing inspection procedures, quality training programs for manufacturing, and a process control system.
- The quality control department is responsible for monitoring the quality of the parts received from suppliers. The department is also responsible for auditing the quality of the products finished in the manufacturing line.
- The materials department is responsible for purchasing the highest-quality parts from suppliers.
- The quality control department is responsible for implementing audit programs at supplier sites and putting in place programs with suppliers to remove the need for receiving inspections. These programs should include the in-plant store and the suppliers the store represents.
- The quality control department is responsible for running a quality information system (QIS) to provide feedback about quality problems to manufacturing and to suppliers.
- The quality control department and the field service department are together responsible for monitoring the quality of the products delivered to customers.
- The product development department is responsible for designing quality and reliability into products.
- The field service department is responsible for collecting quality and reliability data concerning the products delivered to customers and supplying that data to the quality control and product development departments.

A TQC program assigns the main responsibility for quality control to the workers building the product. One important step in accomplishing this is to stop using quality control inspectors in the production process. This passes the responsibility for checking their work to the workers themselves.

Before quality control inspectors are taken off the line, the quality control department must set up a process control system that clearly sets and measures quality standards for every work center in the process. The standards must be simple and precise enough that there will be no questions from the workers. Then, after a training program, the quality control inspectors are moved to other jobs.

To maintain some degree of control, the quality control department will continue monitoring the quality of incoming materials and of finished products prior to shipping. The department will accomplish this by establishing procedures for quality audits and, if necessary, by enforcing 100 percent quality inspections. The quality audits must include products provided by the in-plant store, and the enforced 100 percent quality inspections could occur at supplier sites before shipment to the store.

The quality control department should also collect information on products once customers receive them. The objective is to understand the quality levels of the products as perceived by the customers. In practice, the quality control department will pass this information to manufacturing and to technical support groups. Collecting quality data on the products in the field is a job that requires the coordination of the quality control department and the field engineering group. The field engineering staff are the ones who actually collect the information.

Finally, it is important to develop a QIS to collect data on process quality and on final product quality. The main job of a QIS is to provide quick feedback on quality problems to the people in charge of correcting them. Without consistent problem analysis and correction, there will never be any quality improvement, for the first step in correcting a problem is to understand its causes. The QIS makes available that necessary information.

8.2 IMPLEMENTATION OF A TQC PROGRAM

This section outlines the basic steps required to implement a TQC program. In general, the most important rule of TQC is to promptly correct quality problems as they occur, even if that entails a production line stop. Another important rule is to insist that each part conforms to its specifications. Quality should not be subject to compromise. Therefore, if a part conforms to its specifications, it will be accepted; otherwise, it will be rejected. Once the manufacturer's commitment to this policy is made clear to its suppliers, it is amazing how quickly each quality problem receives attention.

On the basis of the previous paragraph one might be inclined to think the manufacturer is going to check each part against its specifications at some particular time. This is generally true in an inspection system. But in a Just-In-Time system, inspection is considered a waste and should be eliminated from the process. Instead, checking a part's conformance requires checking at some point that the part *performs* up to standard. That is when no compromise should be accepted. If a part falls short of expectations, corrective action must be taken immediately.

The worst time to find out that a part is defective is when it is needed. To avoid this situation, a TQC system should be in place before receiving inspections are eliminated. A good TQC system focuses on prevention rather than inspection. The time to produce a product within specifications is during its manufacture.

TQC in the Factory

It can safely be assumed that most workers want to build high-quality products. It can also be assumed that the workers in a factory will have direct control over the quality of the products they produce. In general, producing a first-class product requires more than ordinary dedication to small details. To achieve this dedication, a manufacturer must train workers to look for those details at the time they are manufacturing the product. The workers' main responsibility is to produce a product that conforms entirely to its quality specifications.

The first step in a TQC program is to make clear to workers that they bear the responsibility for building a quality product and that no inspection system involving any other department is going to absolve them from that responsibility. Manufacturing must build the product right the first time, and they must strive to reach the goal of first-time success as soon as possible.

In practice, manufacturing will not reach this goal without the proper support of the engineers in charge of correcting problems involving the process and raw materials. This support should be timely and unconditional and providing it should be the engineer's first priority.

TQC and Process Control

To put in place a process control system, the manufacturer must first define the process itself. In essence, a process is a sequence of manufacturing steps necessary to build a product. This sequence of steps is normally product-related. Under Just-In-Time, these steps must add value to the product; otherwise, the system considers them a waste. For example, the labor invested in an inspection station in the middle of the process is a waste.

The design of a product directly influences the nature of the process. But not every designer keeps this fact in mind during designing. It is strongly recommended that manufacturing engineers work closely with design engineers early. The manufacturing engineer's job is to define the process required to build the product during the designing, not after the release documentation has reached manufacturing.

A TQC program depends heavily on process control, because its goal is prevention rather than inspection. Prevention requires that checkpoints be set up in the process to monitor the quality of products as they go through. If any quality deviations are detected at a checkpoint, then corrective action is immediately undertaken to bring the process back to within specifications.

Below is a list of some key items to consider in defining and implementing a process control system.

1. Define the process during the product's design phase. In general, simplifications in the process should influence the design, not the other way around.

2. Create a flow diagram with a step-by-step representation of the process from beginning to end. Then try to simplify the process by reducing the number of steps it takes to reach the final step.

3. Assign ownership for every one of the process steps. Devise a quality measurement system for the process at every step and set specific goals for assuring high standards. Implement a tracking system to measure the process quality and also a problem-reporting system.

4. Make the process simple, use no buffer inventories between process steps, and carefully study the labor required for execution. Understand the overhead spent on setup times, inspection, and other activities. Try to minimize nonproductive time.

5. Understand the labor associated with every process step and dedicate enough time to the goal of achieving an even line balance.

6. Define the capacity of the process from the beginning, and allow no over- or underproduction in any particular step. For example, a process step might require a machine tool that can build 2,000 parts per hour but the overall process capacity might be only 500 units per hour. Do not build larger lots in that work center because of the additional capacity of the machine.

7. After setting up checkpoints in the process, implement procedures for defective parts replacement and off-process analysis. Allow no debugging in the process, for process debugging will introduce an unknown time factor and limit the predictability of production output.

8. Assign responsibility for quality to the workers building the product. As a general rule, ask workers to inspect the result of the previous process step before they add any value to the product. The workers must report immediately any quality problem they find. If the problem is serious enough, the workers must stop the line and call a quality team into action.

9. Change the attitude of the workers so that they place quality over production output. The workers' first responsibility is to ensure high quality. If a problem develops, they must stop and report it.

10. Never have static quality and labor goals for a process. Once workers have met the current goals, set higher ones. Always strive for perfection in the process. For example, suppose the yield goal of a particular process step is 95 percent and the labor content is twenty minutes. Once workers can consistently achieve both goals, set new goals of 98 percent and fifteen minutes. Keep challenging the workers and the manufacturing engineers.

11. Define the process in simple terms, and make sure every worker fully understands it. Always have the quality steps of the process done first and the labor steps second. Simplify data collection and paperwork. Collecting data and filling out forms are a waste under Just-In-Time.

12. Set up a process control system using any of the common methodologies. Once the system is evaluated, redefine the process controls for optimum results.

TQC Problem Detection and Solution

A TQC program concentrates on prevention and on the permanent removal of problems. Temporary solutions or the indefinite postponement of problem solving will not be accepted. This requires a change of attitude on the part of the production workers and the technical staff responsible for solving problems. The workers must report problems immediately and the engineers must solve them immediately.

The process of detecting a problem and solving it is systematic and easy to follow. The problems impeding the success of such a system are due to people and not to the process itself. The people involved must have the discipline to carry out the procedures for problem detection and solution consistently and with minimum supervision. In short, they must understand and accept the procedures.

The steps for problem detection and solution are simple.

Step 1: Process performance and problem detection. The best way to detect process problems is to measure performance against control limits. For example, in a particular process step, the worker constantly compares the actual yields of the process against the target yields. When the worker misses the target, he or she must report the problem immediately. At that moment, the worker must have clear instructions as to whether or not to stop the production line. The severity of the deviation will determine the decision. In general, even minor deviations should result in a line stop. This policy will produce many line stops, but it will also force the correction of many problems that cause low production rates and waste of material and labor.

Step 2: Problem analysis. The second step in removing a process problem is to analyze the situation systematically. The manufacturing support staff are responsible for the analysis. They must do it off line to avoid interruptions in production. If the problem with the process requires a line stop, then the support staff can do the analysis on line until they decide on a corrective action. In general, the manufacturing engineers are initially responsible for solving a manufacturing problem, but if the problem is due to a design flaw, then the design engineers should become involved as soon as possible.

Problem analysis must be exhaustive and it should not be hastened because the line is stopped. The manufacturer should accept a long line stop if it is necessary to avoid a half-baked solution. Input from workers is essential for the analysis of a manufacturing problem. In many cases, workers have a good idea how to solve a problem they report. After all, they have extensive experience in working with the product and the process.

The general rule is to keep the line down for as long as it takes to find out what is wrong with the process. Then the engineers must put their best efforts into fixing the problem permanently. If they follow this policy, there will be a high probability that the problem will not return.

Step 3: Corrective action and problem prevention. The wrong time to solve a problem is when the problem is interrupting the process. A TQC program strives for prevention rather than reaction. The capability to identify potential problems in the process is essential for avoiding line stops and crises.

A good process control system must have the means to document problems and assign them to someone for correction. Later, quality review meetings must review the progress in solving the problems and participate in their solution. Depending on the nature of the problem, many departments might participate in the meetings. For example, design engineering must review a product design problem with manufacturing, manufacturing engineering, and field service staff. Every group must clearly understand the nature of the problem and the possible solutions, including the impact each solution would have on the completed products still in the factory and the ones already shipped to customers.

After manufacturing implements a corrective action, the process needs to be monitored to see whether the problem recurs. The production workers can use the tracking system that measures line yields and failures. The system should be able to detect and report any recurrence immediately.

Solving Multiple Problems

In practice, problems don't come one at a time. They frequently appear in groups and at the most critical times. To handle such a crisis, the manufacturing staff must prioritize their resources and apply them to the critical problems first.

The following simple four-step procedure is useful when there aren't enough resources to solve all the problems at the same time:

1. Identify the three most major problems in the process.
2. Analyze the problems and develop a corrective action plan to solve them.
3. Execute the corrective action plan.
4. Repeat the process again for the three next most major problems.

Avoid the trap of spreading the technical staff too thin, for then problems will only be fixed temporarily and sooner or later will come back to haunt the production line again.

TQC Teams in the Factory

TQC teams form a critical part of a TQC program. Employees from manufacturing and the quality control and manufacturing support departments must participate as members. TQC teams also can include design engineers and should use field engineering staff to monitor the performance of products in the field.

The primary responsibility of the TQC teams is to support the process and to work with the workers to carry out successive improvements, not only in quality but in yields.

TQC teams are responsible for yield analysis and process deviations from control points. They are also responsible for implementing corrective actions to bring the process back within operational specifications. The ultimate goal of TQC teams is to reach a point where there are zero defects; they should not be satisfied with any other level of defects, no matter how low.

Note the reference to TQC teams—in the plural. There can be many teams in the factory, each responsible for a different product line. Also, TQC teams can work with suppliers to improve the suppliers' quality and yields. TQC teams are like doctors who treat marginal patients to improve their health and allow them to lead productive lives.

Quality Compliance and Line Stop

Quality compliance in the process must be recognized as a top priority by the workers. Of course, in some organizations quality is considered to have lower priority than production output. The quality control department has the responsibility for setting the measurement points that will define the process's quality compliance. The department also must determine what factors are to produce a line stop. Quality control staff should specify the criteria in the simplest possible terms and clearly explain them to those involved. In general, quality control staff should not complicate the system with lots of measurements and paperwork. Any excess would be a waste and would only add confusion and create more problems.

The workers in the production line must have the authority to stop the line if the quality falls below set points. Then, the manufacturing managers must have the authority to keep the line idle until the process engineers solve the problem. There is not much sense in giving the workers the authority to stop the line if the managers don't have the authority to keep it idle until they can implement corrections.

It is important that the criteria to stop the line are not subjective. If they are, there will be no controlling the process. Workers must have a clear set of rules defining when the line should be stopped. In cases of doubt or confusion, the workers must also have the latitude to stop the line instead of producing a product of dubious quality.

Once the line starts again, the workers will carefully monitor the process to see if the problem has really been solved. If there are questions about the results of the corrections, they can stop the line again.

Quality and the Poka-yoke System

Poka-yoke is Japanese for mistakeproofing. The concept of mistakeproofing originated in Japan in the early 'sixties as a means to prevent mistakes in a manufacturing process. It is based on the belief that workers are fallible and that sooner or later they will introduce unintentional defects in a process. Poka-yoke involves the use of simple process steps or of poka-yoke devices to produce 100 percent inspection in a process. Poka-yoke also requires immediate feedback when a defect occurs in the system.

A poka-yoke device has two functions. The first is to produce a 100 percent inspection in the process, preventing any deviation from the process steps. The second is to provide immediate feedback to the worker that a defect has occurred. A poka-yoke system involves the methodic use of poka-yoke devices throughout the process to detect defects.

In general, there are two types of poka-yoke systems. One applies control to the process. This system has the capability of stopping the process if it detects a defect. This type of system requires immediate corrective action before the process can continue. The second type of poka-yoke system operates in a warning mode. When a poka-yoke device detects a defect, it immediately triggers a warning device, signaling the problem. The system uses alarms and lights to warn the workers that something has gone wrong.

Poka-yoke devices are generally mechanical in nature. Manufacturing engineers add them to machine tools or sensitive places in the process to monitor any product deviation. Poka-yoke process steps are foolproof operations added to the process with the purpose of avoiding mistakes by workers. Widespread use of both methods constitutes a poka-yoke system.

Typical examples of poka-yoke devices include limit switches, photoelectric sensors, proximity switches, and torque detector switches. A simple poka-yoke process step might be as follows. Suppose a worker has to put two assemblies together with ten screws. The worker fetches the screws from a bin and sometimes gets less than ten. To prevent this, an intermediary poka-yoke step is added: The worker has to count ten screws before starting the assembly operation and put the screws in a secondary container. At the end of the assembly operation, the worker must check to make sure that there are no screws left in the container.

Poka-yoke devices are very effective and work well with a workers' on-line inspection system. The combination of foolproofing and self-checking is much more effective than any statistical quality control program. The double inspection system provides a 100 percent check and makes possible the goal of producing high quality products.

Field Quality Feedback and Problem Correction

When a manufacturer delivers a product, the level of the customer's satisfaction will override any previous quality measurement that the manufacturer had in place. The customer is the final judge of the product's quality. This means that products that pass through the manufacturer's process with a 100 percent quality conformance will not be acceptable if the customers are not satisfied. Conversely, the main goal of a quality control system is to produce products that satisfy customers completely. Thus a quality feedback system to measure the level of customer satisfaction must be implemented. Field engineering is responsible for collecting and sending data on customers to the quality and engineering support groups. Below is a discussion of three key quality measurements that are directly correlated with the level of customer satisfaction.

Dead-on-arrival rates. The first quality measurement at customer sites is the product's dead-on-arrival (DOA) rate. The DOA rate is the percentage of products that don't work after delivery. In general, DOA relates to the concept of plug and play. If the product works to the specifications when the customer plugs in the unit and turns on the power, then the product is functional. If there is a defect or a nonconformance, then that product is DOA.

Some companies consider a product DOA only if it doesn't work electrically. For example, if a manual is missing but the product works, then there is no DOA account. A TQC program under a Just-In-Time system considers any nonconformance a DOA (e.g., if the manual is missing, the power cord is not in the box, or the color of the product is not right). A TQC program must strive for a zero DOA rate.

One of the first steps in a TQC program is to measure the current DOA rate for the product and to come up with a plan to cut the rate in half. Then, once this goal is reached, the goal of cutting the new rate in half should be set. The idea is to strive for the ultimate goal: zero DOAs.

Infant mortality. The second quality measurement at customer sites is the infant mortality of the product. Nothing could be more annoying to a customer than a product that works for a few days and then fails. Infant mortality is the failure rate of a product within a short period of time after delivery. For example, infant mortality might be defined as the failure rate of a product in the first thirty days of use.

Infant mortality failures are usually an indication of poor component quality or inadequate manufacturing testing. Decreasing an infant mortality rate requires not only process improvements but also component and design changes. The steps for reducing infant mortality rates are the same as for DOA rates. In fact, a good program to reduce a DOA rate must also include steps to reduce the infant mortality rate.

Mean time between failures. The third quality measurement of a product is the mean time between failures (MTBF). The MTBF of a product is the average number of hours the product operates before a failure. The MTBF is directly correlated to the reliability of the product and depends on the individual failure rate of its components. This fact forces a manufacturer to consider MTBF goals when the product is being designed and to use components which will produce the desired results.

The MTBF of a product can be calculated by two different methods. The first is theoretical. In this method, reliability engineers calculate the MTBF by developing a statistical model of the product and the failure rates of its components. This theoretical MTBF, normally calculated during the design phase, gives the manufacturer an initial idea of the expected failure rate of the product. The second method is empirical. The manufacturer adds all the working hours of products in the field and divides that number by the number of failures. In general, the empirical MTBF closely tracks the theoretical MTBF. If it does not, the manufacturer should take steps to understand the reason for the difference and take corrective action.

Mean time between service calls. If the product is a combination of hardware and software and other services, the MTBF is not an accurate measurement of customer satisfaction. For example, the MTBF will not measure either service calls concerning software problems or operator errors due to poor documentation. A better measure of customer satisfaction is an index called the mean time between service calls (MTBSC). The MTBSC unmasks problems not related to hardware failure but which cause customer dissatisfaction nonetheless. For computer systems, we recommend monitoring the MTBSC rather than the MTBF. Experience shows that the MTBSC is usually about half of the MTBF. For example, equipment with an MTBF of 8,000 hours might have an MTBSC of 4,000 hours.

Under Just-In-Time, the manufacturer's goal should be to have no DOAs and no infant mortality. Also, it must monitor product use and take corrective action to meet the MTBF design goals. If the product is a computer system or another combination of hardware and services, the manufacturer should monitor the MTBSC instead of the MTBF. Just-In-Time considers all field service calls and repairs in excess of the equipment's normal wear-out cycle to be a waste.

8.3 TQC AND SUPPLIERS

A Just-In-Time system cannot exist without a TQC program that includes suppliers. Just-In-Time deliveries and the lack of buffer inventories require that the manufacturer receive high-quality parts to keep the production process going without interruptions. The need for such parts is increased as a result of the Just-In-Time requirement that receiving and in-process inspections be ended.

The first step in starting a TQC supplier program is to choose the initial set of suppliers. The manufacturer can do this by ranking all its suppliers by the dollar amount and volume of products received. Once this ranking is done, the manufacturer can proceed to select the most desirable candidates. Obviously, suppliers that are not already having quality problems should be selected.

After the manufacturer has selected the initial group, it should draft a TQC program for each supplier, outlining the requirements for supplier TQC certification. Later on, after a supplier has met all the requirements, the manufacturer can start accepting Just-In-Time deliveries without in-house inspections. In return, the supplier will receive certain benefits, including a long-term purchase commitment from the manufacturer (see Chapter 6).

If a supplier fails to meet the requirements of the TQC program, the manufacturer should immediately start looking for an alternate source. It must warn each supplier of this possibility.

Supplier TQC Programs

A supplier's TQC program should comprise three subprograms: a quality improvement program, a process improvement program, and a lead time reduction program.

Quality improvement program. This program is designed to improve the quality of the supplier's product. The main goals of this program are to eliminate the need for inspections in the manufacturer's production process and to allow the factory to produce high-quality products. The fundamentals of such a program are discussed below.

Process improvement program. Process improvement is critical to upgrading the quality of the supplier's product, for there can be no quality improvement without improving the process used to build the product. See Chapters 3, 4, and 5 for more detailed information.

Lead time reduction program. Lead time reduction allows suppliers to respond quickly to process changes so as to maintain quality at high levels. A supplier whose production line has a lead time of 170 days will correct problems much more slowly than one whose production line has a lead time of 45 days. See Chapters 3, 4, 5, and 6 for discussions of lead time reduction.

Supplier Quality Improvement Programs

A supplier quality improvement program should start in the manufacturer's production process and reach all the way to the supplier's production line. The program might even extend to the suppliers of the supplier. The program depends on sharing information, on technical help in process control, and on friendly relations. The key to the success of the program is cooperation—and then more cooperation.

It is helpful to start with the understanding that suppliers are not any happier than customers when their products are of poor quality. The way the manufacturer passes on information about quality will help determine the supplier's degree of interest in fixing outstanding problems. The data must be precise, clear, and very timely.

A supplier quality program should begin by collecting information about the quality level of the supplier's product. The quality control department is responsible for this task, which includes receiving inspections, process monitoring, and field installation monitoring. Quality engineers should correlate and prioritize the information, making it simple to understand.

The next step is to present the information to the supplier. This is similar to presenting a report card. The report should tell what the supplier is doing well and what it is doing wrong. Normally, a manufacturer with a QIS in place should have several months of the supplier's data. This information will be helpful in showing quality trends. If a manufacturer doesn't have a QIS, then it must develop one at the time it starts collecting data about the supplier's quality performance. Telling the supplier about quality is not effective; the manufacturer must have enough data to back up its claims.

The manufacturer should not send the information via the supplier's salesperson or mail it. A face-to-face meeting with those responsible for correcting the supplier's quality problems is preferable by far. The manufacturer's engineers should attend the meeting. During the meeting, the manufacturer should get a commitment from the supplier to dedicate enough resources to review the problems and correct them. Also, it should notify the supplier that it wants to do its incoming inspections on the supplier's premises. Instituting a source inspection program will show the supplier that the manufacturer is serious about working with it to fix the problems.

Source inspection as an intermediate step. Just-In-Time considers not only receiving inspections but also the process of returning defective parts to suppliers to be a waste. Even more of a waste is a line down resulting from a defective part.

Source inspection is an intermediary step on the way to eliminating all inspection of parts. It is an effective way to expose quality problems before the parts are delivered to the manufacturer. In a receiving inspection system, once the supplier ships a part to a customer, the supplier has transacted a sale that includes the defective part. If that part fails to pass the manufacturer's receiving inspection, the customer still owns the defective part and its only recourse is to return the part to the supplier for repair or for credit.

When a source inspector rejects a part, the supplier cannot execute a sale until the bad part is replaced with a good one. And if the manufacturer insists on complete lots of parts, it will put further pressure on the supplier to build a quality product.

Another benefit of source inspection is that the manufacturer saves time in reporting problems. The system is very efficient in showing the supplier quality problems with a part, for the inspector is on the supplier's premises and can point out the problem to the people in charge of correcting it. This is more effective than sending a written report that could be subject to different interpretations.

Source inspection also saves the time spent in returning rejected parts to the supplier. Processing returns demands additional labor from the quality, purchasing, and shipping departments but adds no value to the products.

One of the drawbacks of source inspection is that the inspectors must travel to the supplier's site. This costs money and time. In general, the manufacturer will recoup some of the cost by having lower inventory levels, for there is no need to cover for rejected parts. Also, source inspection will reduce the workload in the quality, purchasing, and shipping departments.

One way to reduce the traveling expenses for a source inspection program is to use contract inspectors in the towns where the suppliers operate. For example, the manufacturer could hire retired inspectors who would like to work part-time and would welcome the extra income. Metaphor had very good success using this approach. The system is effective and requires low overhead. It is recommended that a good training program be developed to familiarize the inspectors with the products before they start working with the suppliers.

Source inspection is very effective in uncovering quality problems with parts before they leave the supplier's shipping dock. The goal is to detect problems before the manufacturer invests time and labor in the parts. Source inspection is able to achieve that goal.

Just-In-Time considers source inspection only as a transitional phase between receiving inspection and the elimination of all inspection. The main goal of Just-In-Time is not to shift quality inspection from one place to another but to increase the quality levels so there is no need for inspection. Source inspection helps to accomplish this transition by revealing quality problems effectively and efficiently, resulting in a minimum of wasted labor and overhead.

Working with suppliers to eliminate receiving inspection. Traditional purchasing departments evaluate suppliers in terms of pricing, quality, and delivery in that order. In a Just-In-Time system, suppliers are evaluated in terms of quality and on-time delivery and then pricing.

Quality improvement doesn't occur overnight. It demands constant dedication and attention to small details and requires many small steps to obtain big results. The goal is to consistently get high-quality parts from suppliers, which would eliminate the need for inspecting the parts when they arrive at the back door of the factory.

A system with no receiving inspection is not easy to achieve; it demands time and hard work. Therefore, it is advisable not to send inspectors home or move them to other jobs until a plan is in place and the results are consistently satisfactory over an extended period of time. The first step is replace receiving inspection with source inspection. When the suppliers are meeting the quality goals, the manufacturer can reduce source inspection to an audit and then finally eliminate it altogether.

Stopping inspection, source or incoming, is a big event for the supplier. The supplier is formally TQC–Just-In-Time certified and enters a very exclusive group of suppliers receiving the full benefits of the program. One way to mark this achievement is to issue a certificate to the supplier in a form a plaque. The supplier then can post this plaque, and the manufacturer can reciprocate by posting a duplicate plaque in its lobby. Figure 8.1 shows a plaque issued to Hamilton/Avnet by Metaphor Computer Systems.

TQC–Just-In-Time certification is not necessarily permanent. The supplier must consistently meet the goals of the program; otherwise, the certification can be revoked. The recall process requires a plan similar to the one for qualifying a new supplier into the Just-In-Time program.

The last step before issuing a TQC–Just-In-Time certificate is to discover what process control measures the supplier is implementing to avoid the recurrence of quality problems corrected during the TQC program. These measures must be aimed at providing zero-defect products for the manufacturer's production line—a Just-In-Time goal.

Supplier information systems. A supplier information system has slightly different roles during the transition period and during the postcertification period of a TQC–Just-In-Time program. During the transition period, it is used to track and report quality problems. Once the supplier has been certified, it functions as an early warning system to detect any deviations in quality.

Figure 8.2 shows a quality report issued to a supplier by Metaphor, where a Pareto list is used to prioritize the defects causing rejects. The quality control department at Metaphor collects this information and then mails it to the supplier's quality control department on a regular basis. Metaphor expects the supplier to respond with a plan describing the actions taken to correct the problems in order of priority. Slow response or apathy on the part of the supplier will result in an immediate suspension of TQC–Just-In-Time certification.

A quality reporting system should be simple and effective, and it should direct the information to those who can correct the problems. The supplier's employees who receive the reports must have the authority to take immediate steps to correct problems. It is strongly recommended that quality reports not be sent to suppliers through the manufacturer's purchasing staff or the supplier's sales staff. Communication lines must be short and effective. That means they should directly connect the manufacturer's and the supplier's quality control departments.

Supplier technical support and corrective actions. In order for a TQC–Just-In-Time supplier program to be successful, technical problems in the supplier's production lines must be able to be addressed and solved. In a non-Just-In-Time environment, this

Figure 8.1 Metaphor Computer Systems TQC–Just-In-Time certification plaque.

responsibility belongs entirely to the supplier. This is not the case in a Just-In-Time system, for the manufacturer demands from the supplier perfect quality and is committed to help achieve this goal.

One way to cooperate with the supplier is to develop technical rapport. This mean that the manufacturer's manufacturing engineers cooperate with the supplier's engineers in defining the best process for the product. This cooperation is especially critical when the supplier is developing a custom product for the manufacturer. In this case, the manufacturer must clearly define the environment in which it is going to use the product. It should then help the supplier in order to ensure the highest level of quality.

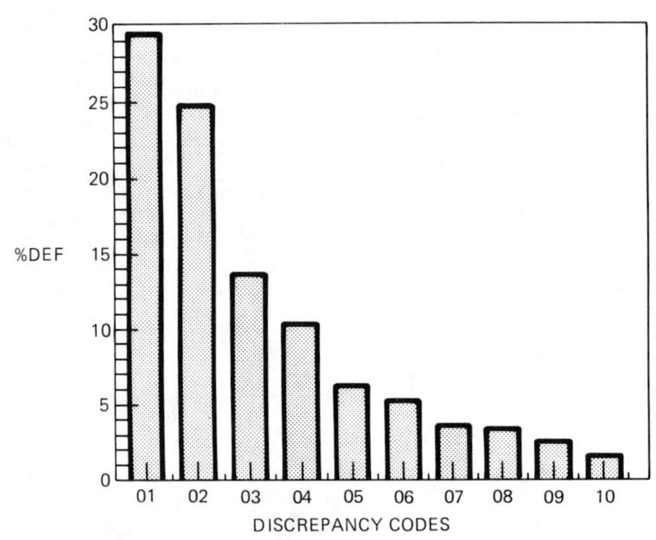

Codes	Description	Qty	% Def of Total
01	Flyback Transformer Interferes W/Plastic Enclosure	109	29.5
02	Picture (ie: Lines, No Video, Interlace)	91	24.7
03	Adjustments (ie: Height, Width Contrast)	50	13.6
04	Color (ie: Green Hue)	38	10.3
05	Physical (ie: Scratched, Broken, Missing or Wrong Part)	23	6.2
06	Down Rev (ie: Eng Prototype, Needs Update)	19	5.2
07	Electrical (ie: Voltage, Peaking, Shock Test)	13	3.5
08	Bad CRT	12	3.3
09	Dead	9	2.4
10	Noisy	5	1.5
	Total	369	

Figure 8.2 Example of the quality report Metaphor Computer Systems sends to suppliers.

The manufacturing engineers and the engineering design staff are the best qualified to convey technical requirements to the supplier. In practice, the manufacturer should not hesitate to use its engineers to work with the supplier's technical staff to correct quality problems. This will bring both companies together and will provide a rich supply of knowledge for the development of future products.

Once the supplier has acted to correct a problem with a part, the most effective way to test the solution is to use the fixed part in the manufacturer's production line.

8.4 TQC PROGRAM OPERATION MODES

A TQC program involves two operation modes. The first is the information mode, in which the supplier receives up-to-date information about the quality of its product. This information should include process yields, field infant mortality, and reliability. The second is the correction mode. If the supplier's quality is below standard, the manufacturer enters into a corrective program with the supplier. This mode is critical, and the manufacturer should not end the program until the supplier's quality again meets agreed criteria.

The TQC program normally operates in the information mode. It can switch to the correction mode if the supplier's quality level falls below standard. The simplest way to determine the operation mode is to agree to a specific quality threshold. For example, if a supplier has a 99.5 percent or better quality yield in the manufacturer's process, the information mode will be used. If the yield falls below 99.5 percent, then the manufacturer switches to the correction mode.

The sequences of events that occur during the information and correction modes are shown below. The manufacturer should not switch from one mode to another hastily and without the proper information at hand. Also, once in a correction mode, the supplier has to prove that it has fixed the problem before the manufacturer returns to the information mode.

INFORMATION MODE

1. List of suppliers in JIT program
2. In-house quality Reports
   - line yields
   - field repair center
3. Field quality Reports
   - Infant mortality
   - Reliability
4. Reports to suppliers and purchasing department
5. Quarterly review meeting with each supplier

CORRECTION MODE

1. List of problems to supplier
2. Supplier commitment to solve problems
3. Technical support to supplier
4. 30 days no response supplier disqualified
5. Temporary source inspection
6. No quality improvements—supplier disqualified
7. Problems corrected—supplier moved to information mode

8.5 TQC AND QUALITY INFORMATION SYSTEMS

Many polished, well-paid, quality control managers like to produce fancy reports and charts. Then they show them only to top management. This is the wrong way to fix quality problems.

A TQC system is useless without a quality information system (QIS). But a QIS will accomplish nothing if the reports only reach high-level management.

The point is to have effective reports and to get them to those who need them most: the workers responsible for solving the problems. In a Just-In-Time system, collecting information doesn't add any value to the product; therefore, it must be minimized.

8.6 QUALITY RESPONSIBILITY IN A TQC PROGRAM

Quality is everybody's responsibility. A quality system becomes successful when all members of the system share the responsibility, including many departments that have nothing to do with the actual process of building the product. For example, let's assume that manufacturing delivers an excellent quality product to a customer. Then the field engineer installing it makes a mistake and the product doesn't work when the customer wants to use it. The customer will think that the quality of the product is poor.

Similarly, if the manufacturer fails to ship operating manuals or it bills the customer incorrectly, the customers' perceptions of the manufacturer will be distorted.

A TQC system crosses many boundaries within the organization. The concept of TQC teams should involve the relevant departments no matter what their reporting lines are. The common objective is to deliver a product that performs as the customers expect it to. Setting those expectations is the responsibility of the sales force. They also have to keep in mind the requirements of their jobs and work to set the right expectations for the product.

Several concepts vital to producing quality products are reviewed below. They should be applied without compromise throughout the company. The quest for quality is constant and never-ending.

The TQC Program and Product Design

There is no better time to start a TQC program for a product than during the design phase. It is then that engineers can analyze the design tradeoffs and make choices that increase the quality of the product. One way engineers can make a substantial contribution to quality is by using parts from suppliers with a proven track record regarding quality.

The Just-In-Time system calls for single-source suppliers and the delivery of high-quality parts. Selecting the right suppliers at the time the engineer is designing the product could make all the difference. The selection process is complex, and engineers must work together with the materials and the quality control organizations. The finance department must also be included to ensure that the suppliers are in good financial standing. But the most important factor, even more important than price, is the quality track record of the supplier and its ability to solve quality problems when they appear.

Manufacturing to Build Quality

One of the main things to remember is that a manufacturer can't improve quality merely by inspecting a product. An inspection system only separates the bad products from the good ones. The only way to produce a quality product is to build quality into the product; no inspection system can do that. This principle must be kept in mind by all production workers. They must always build quality into products—with no exceptions.

A TQC–Just-In-Time system offers no other alternative. There are no additional parts to replace the ones that are rejected. If workers build poor quality into products and they are later rejected, production output will decrease. The manufacturer won't have replacement parts to cover the shortages. The message to workers must be clear: Build quality all the time.

Doing Without Quality Control Inspectors

In a Just-In-Time system, the labor invested in inspecting is considered a waste. This activity adds no value to the product, and the waste is never more visible than in a production line. First, workers spend labor assembling a product. Then another group of workers check to make certain that the first group did their job correctly. This takes the responsibility to do a quality job away from the production workers.

A simple approach is to tell production workers that they are responsible for building the product right the first time and that no effort will be made to check their work. In practice, a manufacturer cannot ensure quality by simply telling the workers they are responsible. It must train them to be their own inspectors and must provide clear instructions on how to do their jobs. The quality control department and manufacturing engineering have the responsibility for providing instructions and training.

To keep the system honest, the quality control department and the manufacturing engineering group must perform random audits of the product. The result of each audit should be a list of quality problems and a set of recommended solutions. The next step is to meet with the workers and discuss the problems and the possible corrections. It is critical that before any changes are made in the process, the workers are able to give their input. Sometimes a manufacturer thinks it has instituted the best solution only to find out later that the workers had a better one. After all, they are the ones who are doing the job every day.

8.7 MANAGEMENT AND TQC

Management plays a very critical role in the achievement of quality. Managers have the responsibility of supporting the quality control efforts throughout the company. They also are responsible for providing workers with the right tools and training to do a high-quality job. Managers cannot ask workers for quality output if they don't provide them with the right process. Managers must be consistent in their directives and not contradict themselves. Mixed messages can be most disruptive. Also, the wrong message will cause demoralization.

When faced with a decision involving quality, the managers must be careful not to compromise. They must always decide in favor of quality. Nothing can be more convincing to the company as a whole than an uncompromising policy concerning quality. The strongest message managers can send the workers is to show their commitment to quality by example, not by words.

Quality Throughout the Company

Quality involves attention to small details. Quality also involves team efforts. No department in a company can achieve the production of excellent-quality products by itself.

Quality is transacting business in a professional manner. Quality is paying bills properly and answering customers' questions promptly and correctly. Quality is fast and effective repairs. Quality is shipping what the customer ordered and it is also equipment that works. Quality is excellence in performing all the activities of a business.

Just-In-Time forces a company to maintain high-quality standards. Conversely, a company cannot have a Just-In-Time system without a TQC system in place. The systems need each other and will not work independently.

TQC and Zero Defects

Over the years there have been many programs to focus attention of workers on quality. Quality circles, zero defects programs, and many other programs have been used successfully in some companies and not in others. Among the main reasons for the failures has been the lack of total manager and worker commitment and the lack of understanding that a quality program must never end. In practice, managers and workers must think of quality as a patient in a hospital intensive care unit and of themselves as doctors who must monitor the vital signs and take corrective action immediately.

In a zero defects program, a company makes the commitment to produce flawless parts. Many companies consider this goal unreachable, but they embark on the program hoping that the mere effort to reach perfection will result in improvement. Other companies only pay lip service to the concept.

A TQC program is a step beyond a zero defects program, for the commitment is not only to build products of excellent quality but also to realign the responsibilities for quality. A TQC program also decreases waste in the process by eliminating receiving and line inspections.

8.8 SUMMARY

The implementation of a TQC system is vitally important for the success of a Just-In-Time system. It is recommended that both systems be implemented at the same time. Also, a manufacturer must take a hard look at its suppliers' quality before accepting them as Just-In-Time suppliers. The suggested approach to increasing quality is to proceed in a series of successive steps and keep pursuing small improvements. Never assume that because the quality is good, the company is out of trouble. Everything could turn bad overnight, and it might take a lot of work and dedication to get the problems corrected and the process back to normal.

It is important not to institute TQC programs with all the suppliers at the same time. Define the program and pick the suppliers in order of importance or quality-related criteria. Then as the company gains experience, it can institute programs with other suppliers.

REFERENCES

BERGER, ROGER W., and THOMAS H. HART. *Statistical Process Control: A Guide for Implementation.* New York: Marcel Dekker, 1986.

CHUNG, CHEN-HUA. "Quality Control Sampling Plans under Zero Inventories: An Alternative Method." *Production and Inventory Management* (American Production and Inventory Control Society) 28, no. 2 (1987): 37–41.

CROSBY, LEON B. "The Just-In-Time Manufacturing Process: Control of Quality and Quantity." *Production and Inventory Management* (American Production and Inventory Control Society) 25, no. 4 (1984): 21–33.

CROSBY, PHILIP B. *Quality Is Free.* New York: New American Library, 1980.

———. *Quality without Tears.* New York: NAL (New American Library), 1985.

DEMING, W. EDWARDS. *Out of the Crisis.* Cambridge, Mass.: MIT Press, 1986.

FEIGENBAUM, ARMAND V. *Total Quality Control.* New York: McGraw-Hill, 1983.

HALL, ROBERT W. *Attaining Manufacturing Excellence.* Homewood, Ill.: Dow Jones-Irwin, 1987.

HARRINGTON, H. JAMES. *Poor-Quality Cost.* Milwaukee: ASQC Quality Press, 1987.

HENDRICK, THOMAS E. "The Pre-JIT/TQC Audit: First Step of the Journey." *Production and Inventory Management* (American Production and Inventory Control Society) 28, no. 2 (1987): 132–142.

ISHIKAWA, KAORU. *What Is Total Quality Control? The Japanese Way.* Englewood Cliffs, N.J.: Prentice-Hall, 1985.

LAFORD, RICHARD J. *Ship-to-Stock: An Alternative to Incoming Inspection.* Milwaukee: ASQC Quality Press, 1986.

LEHRER, ROBERT N. *Participative Productivity and Quality of Work Life.* Englewood Cliffs, N.J.: Prentice-Hall, 1982.

LLOYD, DAVID K., and MYRON LIPOW. *Reliability: Management, Methods, and Mathematics.* 2d ed. Milwaukee: ASQC Quality Press, 1984.

OAKLAND, JOHN S. *Statistical Process Control: A Practical Guide.* New York: Halsted Press, 1986.

OTT, ELLIS R. *Process Quality Control: Troubleshooting and Interpretation of Data.* New York: McGraw-Hill, 1975.

SHINGO, SHIGEO. *Zero Quality Control: Source Inspection and the Poka-yoke System.* Cambridge, Mass.: Productivity Press, 1986.

SUZAKI, KIYOSHI. "A Comparative Study of JIT/TQC Activities in Japanese and Western Companies." Paper presented at the First World Congress of Production and Inventory Control, Vienna, Austria, 1985.

9 ★

Cost Accounting and Just-In-Time

Just-In-Time is usually thought about in connection with manufacturing, inventories, suppliers, and quality control but rarely in connection with the cost accounting and the accounts payable departments. In practice, Just-In-Time affects these departments as much as it does the others, for the operational changes introduced in manufacturing can increase the workload in cost accounting and accounts payable if changes aren't made in their working procedures. Many companies use complex cost accounting systems that take considerable overhead to run. Large amounts of data and numerous reports usually burden those systems and produce very little in return. This excess, in a Just-In-Time system, is a waste.

This chapter reviews the impact that the elimination of work orders and the use of repetitive manufacturing will have on cost accounting. It also will review the impact that frequent deliveries have on the accounts payable department. Suggestions will be made concerning invoicing procedure modifications that can reduce the workload needed to support such a system.

The purpose is to simplify cost accounting, avoid the handling of excess paperwork and information, and make efficient use of the labor required to get the job done.

9.1 JUST-IN-TIME AND PAPERWORK

Just-In-Time crusades against waste, and paperwork is one of the most common causes of waste in any organization. Filing nonessential reports, tracking what is not necessary, mailing copies of papers to a lengthy distribution list—all these activities cause waste and make a simple job complex.

A common trap is to overuse computers to track irrelevant data. There is a direct correlation between the amount of information stored in the computer and the number of employees needed to input the data. There is also a correlation between the number of employees and the number of reports printed and distributed—reports that most recipients don't read, use, or really need.

An effective Just-In-Time system must address paperwork abuses. It also must come up with savings in the way information is used in general. A company would do well to come up with a plan to address this problem in the areas causing most of the paperwork load.

The first step in a paperwork reduction program is to determine what information is really needed and to devise simple ways to collect, store, and distribute that information. In the following sections, the problem of excess paperwork is discussed as it pertains to some of the departments most affected by the Just-In-Time system.

9.2 A COST ACCOUNTING SYSTEM UNDER JUST-IN-TIME

The main job of the cost accounting system is to keep track of the actual cost of products built in manufacturing. This actual cost affects the profit margins of products sold and directly impacts the economic health of the company. A cost accounting department monitors a product's costs by comparing actual costs against standard costs previously stored in the computer's data base.

Three variables compose the actual cost of a product: (1) actual material cost, (2) actual labor used, and (3) actual overhead absorbed. The manufacturing computer system monitors the actual cost of materials by checking the value of the purchase price variances (PPVs). PPVs are the variances against standards of the prices paid at the time the buyers bought the parts. In practice, PPVs can be positive (i.e., the company paid less than the standard cost for the parts) or negative (i.e., the company paid more). Positive PPVs flow through the P/L statements and add to the profit line of the company. The cost accounting department is responsible for monitoring PPVs and reporting them to the purchasing department.

Cost accounting also collects the actual labor invested in building the product and checks it against the labor standards stored on the computer to determine variances. Finally, cost accountants check overhead expenses against absorption rates to see if manufacturing is absorbing them at the rates estimated on the product's standard cost. All this information determines the actual cost of the product shipped to customers.

Another function of cost accounting is to keep track of the inventory in manufacturing. This inventory is normally spread throughout different locations and is at different stages of processing. The cost accounting department tracks the perpetual inventory records in the computer's data base against the physical inventory at hand. It normally implements cycle counting programs regularly by checking random item counts against the perpetual inventory on the computer. Another aspect of inventory tracking is the monitoring of scrap parts in the production line and of excess and obsolete inventory.

Direct Labor Tracking

Cost accounting systems normally use two different methods for tracking a product's labor cost. In practice, the type of manufacturing process usually determines which method is selected, rather than the need to reduce the overhead of the system used. The first method is to use work orders. This method is able to track the product cost in cases in which the company manufactures in discrete quantities or batches. A work order is a paper authorization to build a determinate quantity of the product. The company charges all materials and labor associated with the product built against that work order.

Work orders are cumbersome, difficult to track, and, in time, become a problem to control. Production planners open work orders in the manufacturing computer, collecting all labor and material invested in building the number of units specified in the work order. Then, dividing the total value of the work order against the number of units, they can determine the actual cost of the product and the variances against its standards.

The major problem with work orders is the tracking required to keep the system accurate. Manufacturing organizations have the tendency to keep work orders open indefinitely, which complicates the cost accounting process of calculating actual costs. Manufacturing also splits work orders to keep part of the product moving in the production line. This practice makes the system even more complicated. In a Just-In-Time system, work orders are considered a waste and are avoided. In the following sections, ways to replace the work order system to calculate actual costs will be presented.

The other way to keep track of manufacturing costs is to use a process-tracking system. Manufacturing organizations which do not have batch production lines sometimes use this approach. Process tracking focuses on the rate of products produced and on the consumption rate of raw material and labor required to build those products. The system tracks labor globally by completions instead of associating it with discrete quantities during the processing time of the product.

Process tracking is simpler than using work orders and requires less paperwork. It has evolved into the techniques of repetitive manufacturing, which extends the concepts of process rate and labor capture to the production of discrete products. For example, petrochemical companies have used process tracking to deal with the flow of material in their production lines. The concept of process tracking, in the form of repetitive manufacturing, has also been applied in companies with high-volume production rates. Just-In-Time brought the concept of repetitive manufacturing to a new level; a low-volume production line could still use the techniques to simplify data collection and to accommodate the requirements of a demand pull system.

Work Order Systems and Labor Tracking

Figure 9.1 shows what typically happens when a work order is released to the manufacturing floor. The production planner, using the master schedule as a guide, opens a work order in the computer to build a determinate number of products. The computer lists all the parts

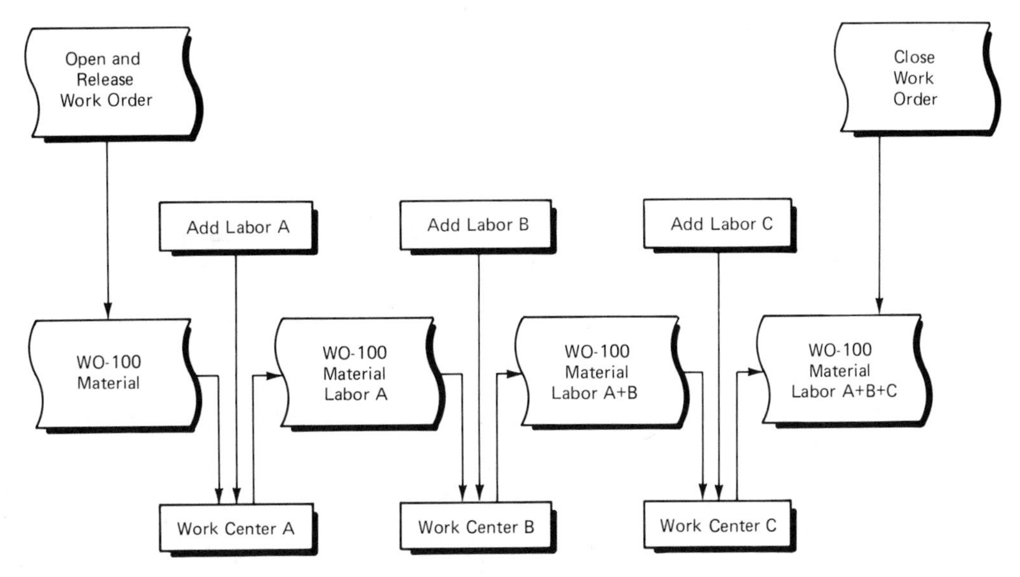

Figure 9.1 Work order processing and completion.

and subassemblies required to build the product, routing the work order to the different work centers where the process is going to take place.

The stockroom delivers the work order paperwork and material to work center A. The workers then execute a task and record their labor on a report card. For sophisticated systems, the workers can enter the labor spent directly into the computer.

After completion, the worker at work center A sends the work order and the partial assemblies to work center B. Then the worker at work center B executes the same procedure, always tracking the actual labor invested in the process. At the end of the routing, the last worker sends the work order and the final product to the stock room. A production planner will close the work order on the computer. The cost accounting department will credit, or charge, any labor variances on the work order against the final cost of the product.

In the previous example, there is clearly an associated overhead in tracking and processing paperwork throughout the process. In practice, the complexity of the paperwork gets compounded if many work orders are released to the work centers at the same time. Workers will have to keep track of the hours invested in each one of them.

Figure 9.2 shows what happens when the production planner has to split the work order to keep some product advancing toward the end of the line. In general, there are many reasons that could hold back part of a work order in a production line. For example, material shortages might allow the building of only part of the work order. Quality problems might also affect yields. The practice of splitting work orders increases the overhead required to track the system even more. It also takes considerable resources in production planning and cost accounting. Split work orders also have the tendency of staying open for longer periods

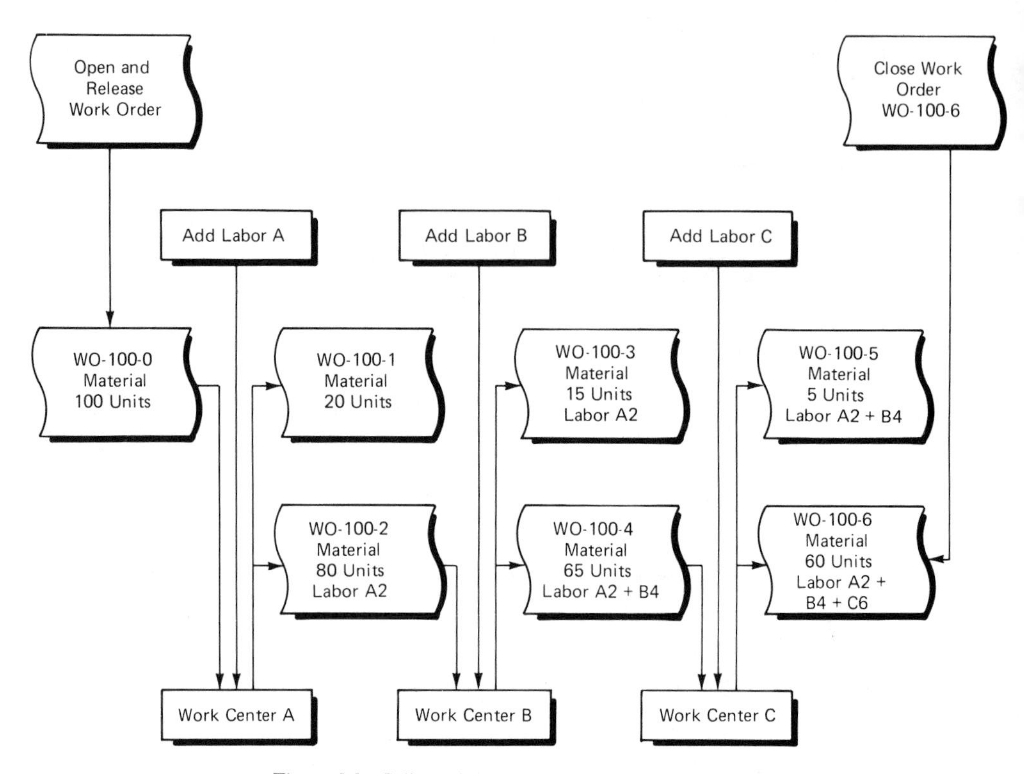

Figure 9.2 Split work orders in a production process.

of time. They become obsolete and nonrepresentative of the real production requirements as planners issue new work orders to compensate for the lack of completion of the old ones.

Figure 9.2 shows the initial work order WO-100-0 issued to work center A; the work order authorizes the building of 100 units of the product. Then, for some reason, the planner splits the work order into two new ones: WO-100-1 and WO-100-2. Work order WO-100-1, with 20 units, stays in work center A for a later completion, and work order WO-100-2 moves to the next work center. The same situation is repeated throughout the process, leaving behind a split work order in every work center. At the end of the process, only 60 percent of the work order has been completed, and the paperwork for one work order has mushroomed into paperwork for six split work orders.

Just-In-Time considers work orders a waste, for the additional overhead in tracking labor and moving paperwork adds no value to the product. Under Just-In-Time, a company can use a simple global tracking of the actual labor invested in a product. It can even refrain from tracking labor as it relates to a particular product, for it can account the direct labor salaries on the overall manufacturing department's expenses. This approach is considered further below. (A labor overhead factor can be applied to the material once the planner releases it to the manufacturing floor.)

Overhead Absorption

Overhead is any expense that doesn't directly contribute to the manufacture of the product (i.e., every expense but direct labor). Overhead includes the rent of the building, the utilities bills, the managers' salaries, the depreciation of the machinery used to build the product, the cost of material scrap, the lack of worker productivity, and the salaries of all the production line support staff.

The cost accounting department keeps track of the overhead expenses and compares them with the direct labor expenses incurred on the production line. The overhead rate is the ratio of the overhead dollars spent by the manufacturing department to the direct labor dollars invested in building a product. A low rate means that the manufacturing department is very efficient and requires relatively little support from the overall organization.

The overhead rate affects the cost of a product. The manufacturer must add overhead expenses to the cost of materials and the cost of labor to determine the total cost of the product. For example, let's assume that two hours of direct labor is spent to build a product. If direct labor has a standard cost of ten dollars per hour, the labor cost needed to build one product will be twenty dollars. Let's assume now that the overhead rate of the department is 300 percent. That means sixty dollars for overhead must be added to the cost of the product. The total cost of manufacturing the product will be the cost of the material plus the labor and the overhead costs. The overhead might not seem very high, but some companies have staggering overhead rates of 1,000 percent.

In a Just-In-Time system, a company should strive for a low overhead rate. This effort translates into less management, fewer support people, less capital equipment, and less rent and utility expenses. One Just-In-Time savings is the reduction of the factory size due to the need for less space for storing material on the manufacturing floor and in the stockrooms. This reduction of space directly translates into less money spent for rent and utilities.

The savings in paperwork translates into fewer staff to input data and to generate and distribute reports. Fewer people means a leaner organization with better communication and responsiveness to changes and problems.

Under Just-In-Time, a company must unburden itself of the deadweight of overhead excesses. The manufacturing department will become very effective in handling its workload and will cost less to the company.

Repetitive Manufacturing and Labor Tracking

Repetitive manufacturing was discussed in detail in Chapter 3. This section concerns labor tracking in a repetitive manufacturing system. The cost accounting department tracks labor globally in a repetitive environment. This means that at the end of a predeterminated period, say a month, the cost accounting department adds all the labor hours invested on the production line. It then divides them by the number of products built. If there are any variances against standards, over or under, it charges or credits them against other costs of manufacturing.

A simple example will illustrate this method. Let's assume that during one month the direct labor salaries, including overtime, come to $36,000. Let's also assume that the amount of product completions during that time was as shown in Table 9.1.

The next step is to compare the actual salaries paid to the direct labor force against the standard salaries they have earned manufacturing the parts. The numbers show that the labor costs include $4,000 unearned. This amount gets charged to the other costs of manufacturing and directly affects the profit line of the company.

To continue the example, let's assume that the overhead rate of the department is 300 percent. Then the overhead absorption of the earned labor is 3 × $32,000, or $96,000. Let's assume that the actual overhead expenses of the department during the month were $100,000. Thus, there was nonabsorbed overhead of $4,000. Cost accounting will also charge this nonabsorbed overhead to the other costs of manufacturing.

The point is that it is possible to compute the actual labor and overhead expenses without having to track labor at the worker level. This simplification yields the same results as a work order system, with the added benefit that workers do not have to spend time filling out time cards and job numbers. The system also saves overhead, since the planners have been freed from the task of opening and closing work orders on the computer and keeping track of them.

One drawback of this system is that global tracking, unlike a work order system, doesn't focus on individual products when workers are not meeting labor standards. For example, the productivity in building one product could be very high and could be compensating for the low productivity in building another. The products taken together will show no problems, for the workers will have completed the total quantity of units according to the build schedule. The solution is to make sure that every worker knows what the rate of production should be (based on the standards). Using repetitive schedules, daily or hourly, makes any missed schedule very obvious.

If the workers consistently miss their goal rates, the manufacturing engineers must work with them to uncover the reasons for the problem. The manufacturing engineers must take corrective action to reach the proper production levels or set new goals and standards.

The global system of tracking labor and overhead is less burdensome than the work order system, but it yields the same results. In some cases, it is even more accurate, because it takes a look at the actual output of the factory. In comparison, the work order system sometimes counts work orders that become split and are never completed. Just-In-Time makes the simplification possible by means of the elimination of wasted time and paperwork.

TABLE 9.1 PRODUCT COMPLETIONS FOR ONE MONTH

Product	Total units built	Standard labor hrs. per unit	Total std. labor hrs. earned	Total $ earned ($10/hr.)
A	2000	1	2000	$20,000
B	50	4	200	2,000
C	500	2	1000	10,000
			3200	$32,000

TABLE 9.2 MONTHLY LABOR AND OVERHEAD EXPENSES

Type of expenses	Amount
Direct labor expenses in salaries	$36,000
Overhead expenses in manufacturing	100,000
Total department expenses	136,000
Labor earned in production	32,000
100% overhead rate applied to labor	32,000
10% overhead rate applied to materials received	70,000
Total earned labor plus overhead	134,000
Unabsorbed labor plus overhead	(2,000)

Overhead Absorption on Materials

There is one further simplification that can help reduce overhead expenses. A company can apply an overhead rate at the time it receives the material in the receiving dock. This approach makes sense, because the material cost of a product is normally much higher than its labor cost. Also, the company spends a large portion of the manufacturing overhead buying that material.

A materials overhead system is simple and easy to use. Let's say a company applies a 10 percent overhead rate to all materials once the receiving clerk enters them in the computer. This charge actually increases the value of the inventory by that amount. The company then applies it against the overhead expenses in the manufacturing organization.

In the example in the previous section, rather than applying a 300 percent overhead rate to labor, the company could apply only a 100 percent overhead rate. Then it could apply a 10 percent overhead rate to materials received during the month. For example, if it received $700,000 in materials during the month and it applied a 10 percent overhead rate to those materials, the actual value recorded in the books would be $770,000.

Table 9.2 shows how the charges will appear at the end of the month.

This method will yield the same results as applying only labor overhead to the department, but it is fairer because the organization spends most of its overhead bringing in materials that are not immediately turned into products.

Labor Absorption on Material Transactions

For a further simplification of the system, a company can look at the expenses in manufacturing as a unit and refrain from tracking direct labor against production completions. Applying an overhead rate to the materials consumed in the production line can be sufficient to absorb the direct labor invested in the products. The point to reinforce is the difference

TABLE 9.3 MONTHLY EXPENSES WITH MATERIAL ABSORPTION

Type of expenses	Amount
Total manufacturing department expenses	$136,000
10% overhead rate applied to materials received ($700,000)	70,000
10% overhead rate applied to materials consumed in the production process to build the product ($640,000)	64,000
Total manufacturing absorption	134,000
Unabsorbed labor plus overhead	(2,000)

between overhead applied to the materials upon reception and overhead applied to the materials released to the production line to build the product.

Table 9.3 concerns the same example as before. The assumption is that the company received $700,000 in materials and it used $640,000 in the production process to meet the month's production schedule.

The example illustrates a simplified system of absorbing manufacturing expenses. The actual percentages applied to the material received and consumed would vary depending on the organization and the type of product. The system is simple to track and requires very little overhead to operate. It also helps to solve the absorption problem with material-intense products that require very little labor to build. The cost accounting personnel required to keep track in this system is considerably less than in a labor-tracking system. Also, the workers will not waste time accounting for their labor hours and will be freed for more productive tasks.

In practice, this method of absorbing labor and overhead provides only a global view of the organization's performance. The system hides irregularities on the production line. One solution is to set clear production goals at every work center based on the labor standards of the product built there. Tracking this output rate will allow manufacturing managers to spot any problems early enough.

9.3 PAPERLESS PURCHASING

Chapter 6 reviewed the role of the purchasing department in dealing with suppliers under a Just-In-Time system. The topics include long-term contracts, frequent deliveries, and single-source suppliers. This section is concerned with the effect that a Just-In-Time system has on the purchasing department itself. In general, a Just-In-Time system tends to reduce the workload in purchasing as a result of eliminating multiple suppliers, frequent bids, and continual switching of suppliers. On the other hand, the system tends to increase the workload as a result of increasing the frequency of deliveries and the amount of purchase order tracking. Consequently, the purchasing department needs to change its operating procedures and introduce new concepts not normally used in purchasing.

Another aspect to consider is the simplification of the paperwork inside the department. Purchasing is a paperwork-intensive activity and the department is a natural place to look for waste. Distribution of multiple purchase order copies, frequent filing, and complicated purchase requisition procedures are good targets for simplification.

Purchase Order Routing

Figure 9.3 shows a typical flow for a purchase order. After running MRP, the materials planner uses the information to fill out a planning sheet on every part required to build the product. The planner uses the MRP order points to outline the quantities and the dates for the procurement of the parts. Then the planner fills out a form requesting a purchase order to buy the material. The planner sends this information to purchasing. A buyer responsible for that commodity will request quotes, select the supplier, and place the order. Once the buyer places the purchase order, he or she sends the original to the supplier and distributes copies to the purchasing master file and accounts payable. This procedure leads to excess paperwork and redundant files and produces only wasted time and overhead.

In general, buyers place purchase orders to cover a span of four to twelve months and call for monthly deliveries. Then the buyers monitor the supplier's performance and expedite deliveries in cases where the supplier is not performing. After every running of MRP there is normally a flurry of activity by planners and buyers to review the status of purchase orders and to accommodate any rescheduling that the MRP had requested. All this activity does nothing but increase the paperwork traffic in the department.

Figure 9.4 shows a simplified flow for a purchase system using long-term Just-In-Time contracts. The contracts, eighteen or twenty-four months long, specify the total quantity of product, the price, and the Just-In-Time deliveries and rates. Communication between the planner and the buyer is mainly for the purpose of sending a monthly rolling forecast to the supplier to help fine-tune the next four to six months of production. Then the planner will give the buyer daily or weekly releases based on actual material consumption in the production process.

In the case of hourly or daily deliveries, the buyer must set up with the supplier a fast, reliable way to communicate the number of units in the shipments. A simple system using phone calls, a telex, or a copy of a Kanban sent by a Fax machine will suffice. This system will inform the supplier how many parts to ship in the next shipment. For a steady production rate, it will be necessary only to call the supplier when there is a change in the rate; the agreed current rate is essentially a default quantity. For example, the buyer could have instructed the supplier to ship 200 parts daily and would only call when a different number is desired.

Blanket Purchase Orders

One of the problems with frequent deliveries and rate changes is the additional paperwork required to account for all the transactions. There is no question that purchase orders and

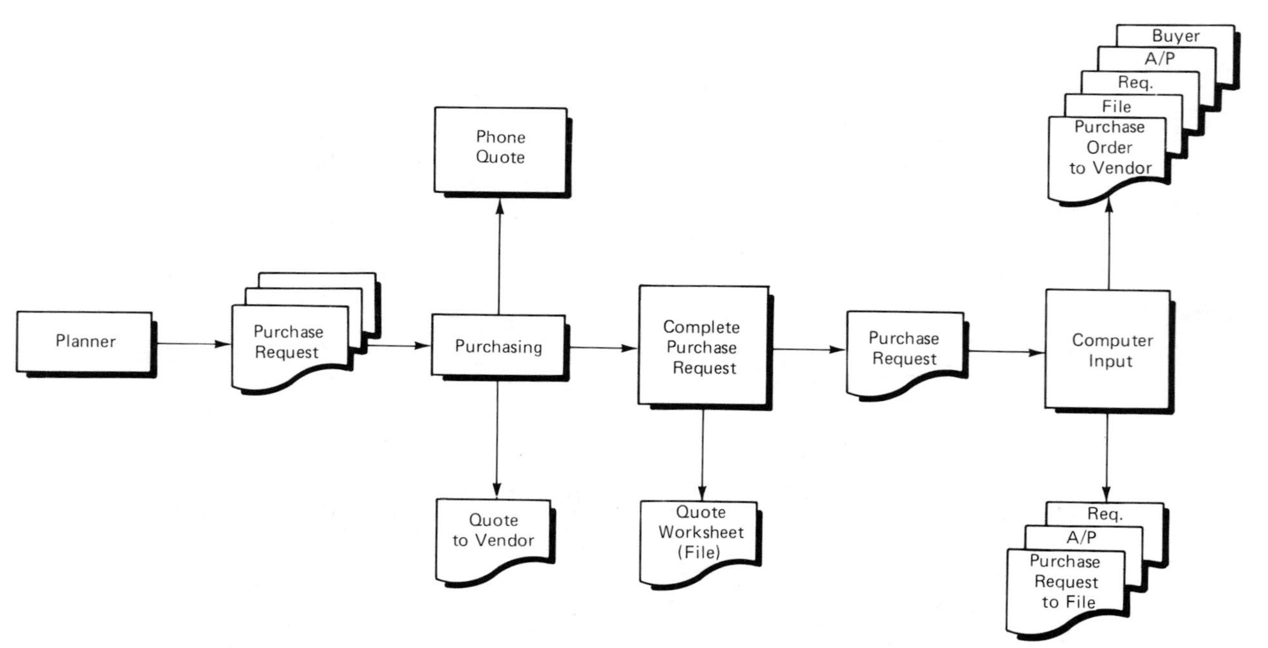

Figure 9.3 Typical flow in a purchase order system.

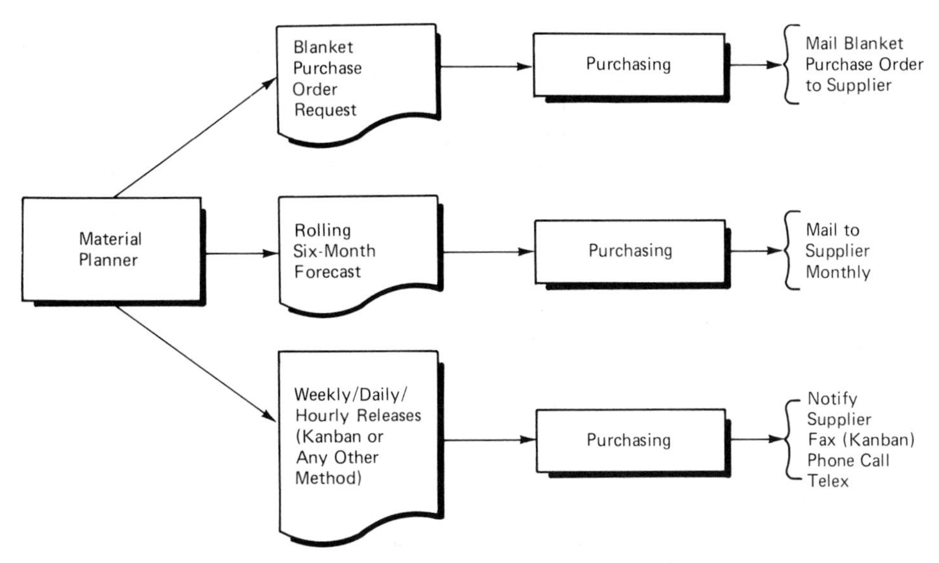

Figure 9.4 Paperwork flow in a Just-In-Time purchasing system.

invoices tracking daily deliveries will be burdensome and prone to mistakes. This in turn will translate into more work and an increase in overhead.

The simplest way to solve this problem is to combine the actual deliveries into monthly statements by issuing a blank purchase order covering the Just-In-Time contract. This purchase order will specify total unit quantity and price, but it will not call for a delivery schedule. When mailing the purchase order to the supplier, the buyer should include the first rolling forecast, detailing deliveries. This will be the first product release against the blank purchase order, and no further paperwork will be required.

Figure 9.5 shows the transactions required to receive parts covered by a blanket purchase order. The receiving clerk inputs into the computer the number of units received and credits the blanket purchase order. Then the clerk mails the original receiver to the accounts payable department and also mails a copy to the buyer. The only possible problem is if the supplier has not shipped the exact quantity of parts. The buyer has the responsibility for

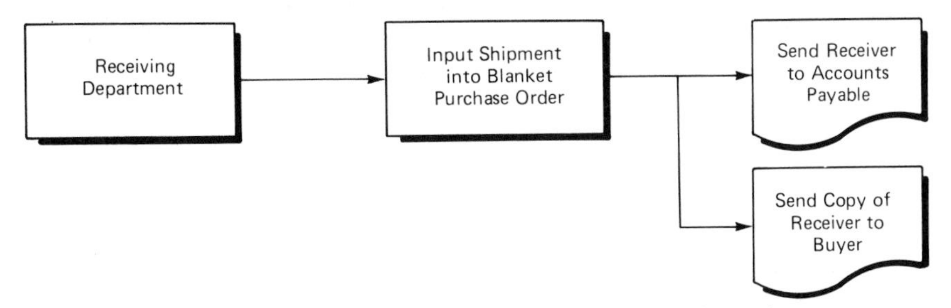

Figure 9.5 Combining transactions into a blanket purchase order.

checking the copy of the receiver and making sure that it matches the quantity he or she previously released. This task will be simple, because the buyer either has the supplier at a constant rate of delivery or has just given a new release to cover for that day's delivery.

Metaphor found that not many suppliers were capable of handling blanket purchase orders and consolidating the frequent deliveries and invoices into a single monthly statement. The most common problem was the inability of computer systems to consolidate invoices into monthly statements. Suppliers of manufacturing software packages are urged to design purchases modules to handle this type of transaction. Just-In-Time procedures leave no other choice but to combine the paperwork from frequent deliveries into single statements.

9.4 ACCOUNTS PAYABLE AND JUST-IN-TIME

The Just-In-Time policy of using small lots and frequent deliveries could severely impact the workload in the accounts payable department. The department will see a marked increase in paperwork activity, for staff will have to match more invoices against receipts, keep track of more shipments, and consolidate invoices into single check runs.

Blanket purchase orders offer a simple solution to this problem, but if the suppliers do not make arrangements to use them, a constant stream of invoices will flood the cost accounting department. For example, assume that a company changes the delivery schedule of a supplier from once a month to a daily rate. This new shipment rate will increase the invoicing activity from 12 to 260 invoices a year. If there are ten suppliers making daily deliveries, the invoice traffic will increase from 120 to 2600 invoices a year.

One way to solve this problem is to ask suppliers not to invoice every shipment but rather to send group invoices covering many shipments. The company might request monthly invoicing of shipments or, if it has a good system, no invoicing at all. The company could pay each supplier on time, based on its receiving records collected during the month. This method could cause trouble if the company doesn't pay its bills on time or the accounting department doesn't keep current records of payments and receipts. It is probably best to ask suppliers for monthly invoices and to pay them accordingly.

One possibility is to set up a system for transferring funds electronically. Then the company can use its computer to transfer funds matching the invoices for a determinate set of deliveries. The computer can automatically set an offset time of thirty days to execute the transfer of funds. This system will eliminate the writing of checks and the associated overhead.

9.5 THE IMPACT OF JUST-IN-TIME ON INVENTORY TRACKING

In considering the impact of Just-In-Time on inventories, it is usual to neglect the reduction in overhead associated with keeping track of the inventories. A smaller inventory reduces the work required to maintain accuracy, for it is easier to avoid mistakes transacting material when there are fewer parts involved in the transaction. For example, less WIP and shorter factory cycle times help to increase accuracy in counting the materials moving through the production line. This improvement in counting reduces the amount of mismatching between the physical inventory and the inventory records on the computer. It also strengthens the effectiveness of the MRP system in calculating the requirements for materials. This translates into a reduction of shortages and last-minute surprises.

Cutting Down on Inventory Transactions

One way to reduce overhead and errors in inventory tracking is to cut down on the number of transactions on the computer when materials move through the factory. In a typical manufacturing organization, inventories are separated into imaginary buckets identified by computer account numbers. When a part moves from one place to another, a computer transaction is executed which subtracts the part from one bucket and credits it to another. In practice, there is a one-to-one correspondence between buckets and physical locations (e.g., the stockroom and WIP, finished goods inventory, and MRB locations). These locations can also be associated with the individuals responsible for transacting the materials they move.

Figure 9.6 shows a typical process flow in a manufacturing plant. The figure also shows the different departments involved in transacting the materials as they travel through the production line. A typical part traveling along the main process path will be subject to six different transactions before it reaches the finished goods inventory. The purpose of so many transactions is to properly account for the materials at any particular location in the process.

One possible simplification of the system is to reduce the number of boundaries that a part has to cross as it moves through the process. Figure 9.7 shows an example of this kind of simplification. Here the entire manufacturing floor is considered to be a single bucket location, which reduces the number of transactions in the main path of the process from six to three.

Further simplification can be achieved by using a wall-to-wall inventory location. The idea is to consider all manufacturing as a single bucket. Figure 9.8 shows a simplified wall-to-wall transaction flow in which the number of transactions is down to two.

The extreme importance of reducing the number of transactions in the factory must be emphasized. Just-In-Time calls for the frequent moving of material in small lots. This entails that workers will be increasingly active in the nonproductive task of transacting materials. Just-In-Time, however, considers this activity a waste. In order to resolve the conflict, the system must be simplified so that there are fewer fences the materials have to cross as they progress through the process. This requires that the number of inventory buckets storing materials must be reduced.

Physical Inventory and Just-In-Time

Most companies have an annual physical inventory before closing the fiscal year. Others may find a need to have physical inventories several times a year. Just-In-Time helps the physical inventory process in two ways. First, smaller inventories in the factory reduce the

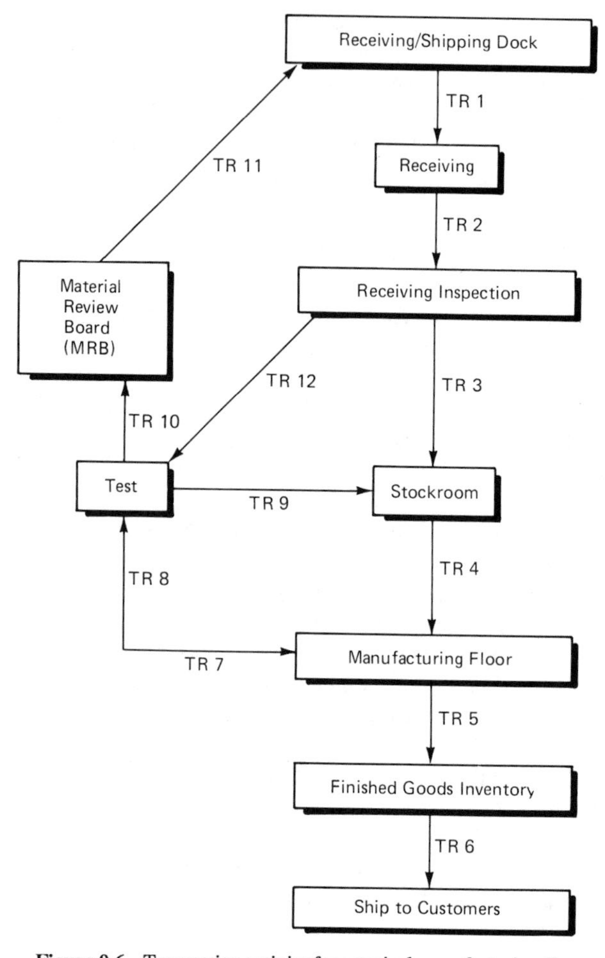

Figure 9.6 Transaction activity for a typical manufacturing flow.

time it takes to count them. Daily schedules and small lots lower the WIP inventory, simplifying the physical materials count. Also, having less material on hand improves the accuracy of each count.

Second, Just-In-Time helps in the transaction process. Reducing the number of transactions in the factory decreases the probability of human error. It also makes it easier to reconcile the transactions when a discrepancy occurs. But all this simplification is insufficient without a system of checks and balances, that is, a cycle count program to monitor the accuracy of the inventory.

The cost accounting department is normally responsible for setting up such a program. It is recommended that all those in charge of transacting materials participate in the program

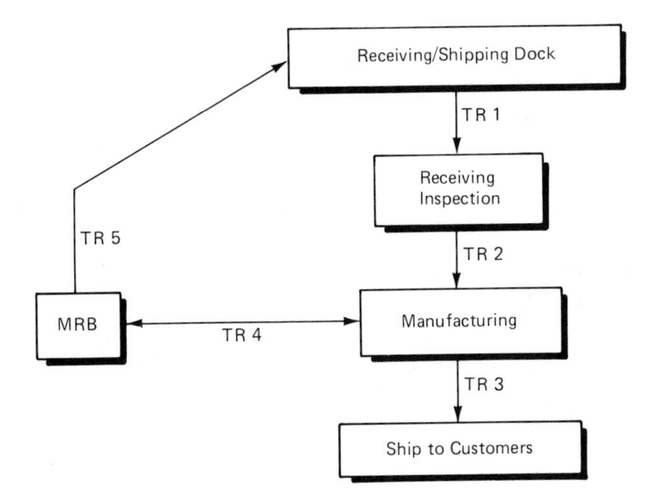

Figure 9.7 Simplified process with a single manufacturing bucket.

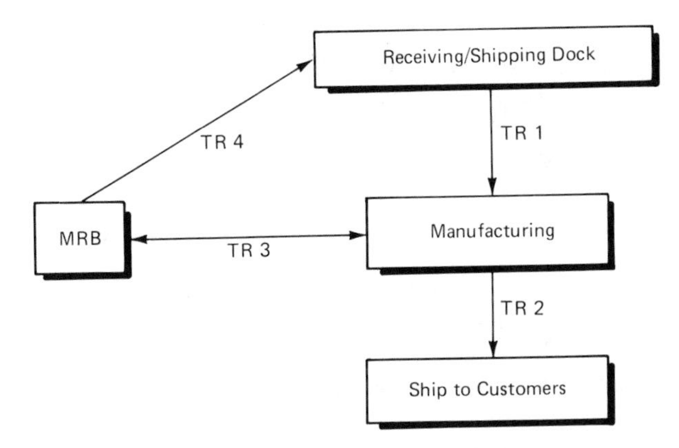

Figure 9.8 Wall-to-wall transaction flow.

and that there be weekly meetings to review the results of the cycle count and any operational issues. These meetings will help to keep the attention focused on the inventory transactions. The process of improvement and attention to inventory accuracy will yield returns many times greater than the labor invested.

9.6 SUMMARY

An effective Just-In-Time program should include the cost accounting department from the beginning. The cost accounting staff should understand the Just-In-Time concepts clearly and should participate in the creation of the procedures used in the system. One point to remember is that there is no teaching of Just-In-Time cost accounting in schools. Some of the new simplifications proposed under Just-In-Time are contrary to the common belief in the necessity of accounting control. The best way to resolve potential conflicts is to educate the cost accounting staff about Just-In-Time concepts. The goal is to make the managers fervent supporters of the new system and to encourage them to make the necessary changes.

As for the purchasing department, not all procedures should be changed overnight. The buyers must first be given proper training in the new system. Perhaps the best strategy is to change the procedures for handling a supplier as it enters the Just-In-Time program. This incremental implementation will allow the purchasing staff to fine-tune their procedures and learn the new process by initially applying it to only a small number of suppliers.

Finally, it is advisable not to start a Just-In-Time program without including these two departments. If they are not ready for the change, the implementation of Just-In-Time is going to become stalled as a result of the additional paperwork load. The staffs will then feel demoralized, for the new system will have created more work, contrary to their expectations. This will increase the chance of failure.

REFERENCES

CANTWELL, JIM. "The How and Why of Cycle Counting: The ABC Method." *Production and Inventory Management* (American Production and Inventory Control Society) 26, no. 2 (1985): 50–54.

CHRISMAN, JAMES J. "Basic Production Techniques for Small Manufacturers: Initial Preparations." *Production and Inventory Management* (American Production and Inventory Control Society) 26, no. 2 (1985): 130–145.

CONSTANZA, JOHN R., and DAVID R. WAGNER. "J-I-T Cost Accounting." In *Just-In-Time Manufacturing Excellence*. Published by authors, 1986.

DAVIS, GINGER M. "Improving Work-in-Process Inventory Accuracy in the Process Industry." *Production and Inventory Management* (American Production and Inventory Control Society) 26, no. 2 (1985): 91–103.

HALL, ROBERT W. "Planning and Measurement of Stockless Production." In *Zero Inventories*. Homewood, Ill.: Dow Jones-Irwin, 1983.

HAMILTON, SCOTT, and SCHROEDER, ROGER. "Computer-based Manufacturing and Accounting Systems for Smaller Manufacturing Firms." *Production and Inventory Management* (American Production and Inventory Control Society) 25, no. 4 (1984): 92–105.

HORNGREN, CHARLES T. *Cost Accounting: A Managerial Emphasis*. Englewood Cliffs, N.J.: Prentice-Hall, 1982.

POWELL, CASH. "Cycle Counting: A System for Reconciling the Differences." *Production and Inventory Management* (American Production and Inventory Control Society) 25, no. 3 (1984): 92–101.

THOMPSON, RICHARD L. "How to Achieve and Maintain Inventory Accuracy." *Production and Inventory Management* (American Production and Inventory Control Society) 26, no. 1 (1985): 38–45.

10 ★

Product Design and Just-In-Time

During the implementation of Just-In-Time, a company will invest much effort in dealing with the factory process and the materials associated with the products. It also will spend extensive time working with the suppliers who account for most of the dollars spent on parts. It will generally pay very little attention, however, to a very critical aspect of the system. Just-In-Time must start at the product design stage rather than wait until the documentation reaches the manufacturing floor. By that time, it will be too late to put in place the things most likely to make the system run smoothly.

The product development engineers can greatly influence the implementation of a Just-In-Time system just by selecting the right parts and suppliers for the new products. They can also structure the documentation so that the material will flow in the factory following Just-In-Time principles. Finally, they can design quality and ease of manufacturing into the product. All these factors are key ingredients of a successful Just-In-Time system.

Another area where product development can contribute to the Just-In-Time system is in the organization of the documentation to reflect the manufacturing process. This includes releasing part specifications that do not place an unnecessary burden on suppliers. The most important areas where design engineers can make a substantial contribution to Just-In-Time are discussed in the following sections.

10.1 TQC AND ENGINEERING

Chapter 8 reviewed the principles of TQC and how they affect many departments in a manufacturing organization. This section reviews the relationship between a TQC system and the product development department.

Many quality experts say that a company cannot inspect quality but rather must build quality. Nothing is truer, but there is one proviso: Quality has its origins in the design phase. If the product development department releases a product designed without quality in mind, there is nothing that manufacturing can do to improve its quality. All that manufacturing can do is to put the product together according to the prints. Then the only way to solve a design problem is to change the design and the prints.

The question is: How do engineers design quality? Any good design engineer can answer this question, for there are many factors that contribute to the quality of the product. In general, a product is a combination of raw materials, a set of design rules, and a documentation set describing how to put the product together.

The first step in producing a good quality product is to select high-quality raw materials. Good quality products are only possible if the individual components are also of good quality. This is the single most important factor in designing quality into a product. In selecting parts, quality must be the highest priority, not second highest behind price. The investment will pay off by easing the manufacturing of the product, producing high yields, and satisfying customers.

The second step is to design the product using safe design rules. In practice, there is a fine line between using worst-case design rules and understanding the limitations of the parts used in the design. The engineer must clearly identify the performance limits of the individual components used in the design. He or she must also know the overall margin requirements for the product. The main goal is to design a product with safe margins which will cover its operational specifications.

Many design engineers neglect the margin issue until the end of the project. It is common to design the product first, test functionality, make corrections, and then test for design margins. Nonetheless, it is advisable to consider quality and margins first, then to design functionality afterwards. The goal is to have quality and design margins influence the final design rather than to complete the design and see what the quality and margins are.

Documentation can also tip the balance and determine whether a product is successful. Documentation is the final output of a product development group. In general, the documentation should specify the product in such a way as to represent clearly the designers' intent. To achieve this goal, development engineers must carefully review the level of information included in the documentation, for problems can arise from providing too much information as well as from providing too little. Lots of irrelevant data will serve no other purpose than to confuse those using the documentation.

One final comment: Design engineering must participate on quality improvement teams and help review manufacturing problems. Their function should be to clarify the difference between design and process-related problems and to explicate the impact of the product design on the process quality. By doing this, they will help the design group to gain manufacturing experience applicable to design of future products.

Rules to Remember during the Design Phase

Following are some important rules that design engineers should keep in mind when designing products for a Just-In-Time system:

1. Evolve new designs on the basis of previous experience. Make full use of knowledge gained from past technologies to develop better products.

2. Specify parts of the highest possible quality in designing the product. Quality must be more important than price.

3. Select suppliers with the best quality track record to supply parts.

4. Reduce the number of parts to the fewest possible. Also, select parts that will require the minimum set of suppliers. If there is an in-plant store in the company, give priority to parts the in-plant store can support.

5. Design all the mechanical assemblies to be foolproof. Make sure there is only one way to assembly each part.

6. Reduce the levels of the bills of material as much as possible. Work with the manufacturing engineers to have the documentation closely represent the process used to build the product.

7. Design for testability. This applies to the individual subassemblies and to the total product. Consult with the manufacturing test engineering group before finalizing the paper design.

8. Set a reliability goal for the product before designing it. Also, set a goal for the product's operational margins. These two goals should influence the design, not the reverse.

9. Consult the field service department to develop a field service strategy for the product before designing it. Outline which units will be field replaceable and which ones field repairable. Develop a spare strategy during the design phase. All these factors will affect customer satisfaction and the amount of money spent in field service.

10. Work with the materials department and the manufacturing engineering department to design the product so it can be built efficiently using repetitive manufacturing and Kanban.

11. Keep in mind during the design phase what type of special tooling is going to be needed to build the product. Try to use equipment already on the production line. A common mistake made during the design phase is to assume that the company can capitalize on new equipment without affecting the profit line. In practice, the company has to depreciate all equipment purchases, which adds to the manufacturing overhead.

12. Develop practical part specifications for suppliers certified under the Just-In-Time program. The specs must be clear and simple to understand and not full of irrelevant details, which will complicate the supplier's process. Try to understand the supplier's process capabilities before writing the specs.

13. For custom built parts and large OEM products, perform on-site surveys before selecting a supplier. The quality, purchasing, and finance departments must participate in the selection.

14. Provide engineering change orders (ECO) quickly to help the manufacturing organization respond immediately to design problems that surface in the production process. Produce simple, easy-to-read documentation that is unburdened by unnecessary information or constraints. Study carefully the effectivity dates of engineering change orders before releasing the paperwork.

15. Track the production yields until the manufacturing organization is far up the learning curve. Try to distinguish production process yield problems from low-yield-margin problems. The latter are clearly design problems.

16. Work with the field service staff to understand the product's field problems. Try to solve any design-related field problem as quickly as possible. Remember that the customer is also aware of the problem and has already paid for the product.

17. Participate actively in quality improvement teams. Assume responsibility for any problem that can possibly be attributed to a design error. Don't be defensive—anybody can make a mistake.

18. Start a cost reduction program for the product before it reaches the manufacturing floor. One way to reduce cost and improve quality is to eliminate parts.

10.2 COST REDUCTION STARTS DURING DESIGN

Many product development departments complete the design of a new product, release the documentation to manufacturing, and then start considering a cost reduction program. In practice, this means that engineers go back to their designs to make improvements they could have done initially. The best time to start a cost reduction program is during design—long before manufacturing receives the documentation.

Just-In-Time calls for a small supplier base. The most effective way to accomplish this is to reduce the number of parts required to build the product. One helpful method is to have a design review team question the design and ask the engineer to try to come up with alternatives in order to eliminate parts. The design review team must make sure that the engineer has analyzed every part used in the design from the point of view of functionality. They also must make sure that when a new part is required, the engineer has seriously tried to use a part available from a current supplier.

At first, this procedure might seem complex and time-consuming. But after a series of design reviews, the final product will cost less, use less parts, require fewer suppliers, and be more suitable for a Just-In-Time release to manufacturing.

In-Plant Store Support

Design engineers can directly support the success of the in-plant store by including in their designs parts the store can support. Once the in-plant store is established, it is to the company's advantage to buy as many parts from it as possible. Every new part bought from the store not only reduces the company's inventory but also relieves the overhead in the materials department. The increase in in-plant store sales will also contribute to the long-term stability of the arrangement.

Selecting parts from the store should be a routine activity during design. The engineers can also make use of the documentation facilities in the store (e.g., they have easy access to part specs and supplier catalogs). An additional service is the procurement of samples. All of these activities will save the engineers time in dealing with salespersons from different suppliers. In short, the design engineers should view the in-plant store as a resource for them as well as for manufacturing.

10.3 ENGINEERING SUPPORT IN THE FACTORY

Just-In-Time forces a manufacturer to face quality problems as soon as they occur, for there is no buffer inventory to cover for the defective parts. This situation should be accepted as something positive. The pressure for urgent action, however, increases the risk that those in charge of solving the problems will only apply band-aids. Therefore, when problems arise, the manufacturer must demand permanent solutions, not quick, temporary ones.

In practice, the arrangements for providing technical support to manufacturing vary depending on a company's management philosophy. Some companies have an engineering support group in charge of solving technical problems with the product. This group reports either to manufacturing or engineering. It has engineering change authority and enough experience to understand the design trade-offs as they relate to manufacturing. One of the main roles of this group is to be a buffer between the design engineers and the laborious job of supporting manufacturing.

The main drawback of this approach is that it prevents design engineers from acquiring direct manufacturing experience, which might help to improve later designs. Also, it does not completely isolate the design engineers from problems, for when a technical problem is very difficult, the support engineers will have no choice but to involve the actual designers.

One problem in having design engineers support manufacturing is that conflicts of priorities can occur. In general, once engineers release a product, they have little motivation to provide manufacturing support. They would rather spend their time designing the next great product. Conversely, their support of manufacturing could be so intense that they might spend most of their time supporting, not designing. This is not a problem in a Just-In-Time system. A new product that requires a large investment in manufacturing support means that the engineers did not complete the design properly before releasing it. In that case, the design engineers would be the best ones to correct the problems.

One last important point is that engineering support should not be confused with manufacturing engineering support. Manufacturing engineers solve process-related problems while engineering support is responsible for design-related problems. The boundary of responsibility is sometimes difficult to determine, for there can be problems related to both process and design. In these cases, both groups should work as a team.

Manufacturing Support Response Time

Regardless of which group supports manufacturing, Just-In-Time demands a faster response to solve problems than in a conventional manufacturing environment. The company must sensitize the support groups to this fact and ask them to always act urgently when a problem occurs on the production line.

Another area that needs streamlining is the processing of engineering changes. For example, a problem that has caused a production line to remain down can't wait for five days to process an engineering change order (ECO). Conversely, support staff should not accept verbal instructions to fix a manufacturing problem. One way to solve this problem is to use emergency ECOs. That is, the ECO clerk hand-carries the ECO paperwork to the members of the engineering change board to expedite its release. This method should yield ECOs in a few hours rather than days.

10.4 ENGINEERING SUPPORT TO SUPPLIERS

Just-In-Time calls for a closer relationship with suppliers. The relationship should go beyond the normal arm's-length business transactions so prevalent in the business world. It calls for a clear understanding of each other's processes and capabilities. It also calls for sharing technical support when a critical situation demands it. A company must keep this in mind when it is selecting Just-In-Time suppliers, for it will eventually commit considerable technical resources in working with the supplier to solve quality problems that could affect the Just-In-Time program.

Product Specifications and Supplier Production Processes

The best thing that design engineers can do to ensure the procurement of a quality part from a supplier is to create clear and simple specifications for the part. If the company can use an off-the-shelf part, a standard supplier's specs will suffice. But if the product is special or deviates from the standard part, clear specs are of vital importance.

Engineers have a tendency to load specs with unnecessary data, which confuses the supplier and increases the cost of the part. They could also go the other way and provide too little information to manufacture the part in a consistent way. It is recommended that before they write the specs for a part, engineers clearly understand the production capabilities of the supplier.

One way to understand the supplier's production capabilities is to have design and manufacturing engineers visit the supplier's production line. Also, the company should invite the supplier to visit its process and study the way it is planning to use the part. Once this interchange has been completed, engineers can specify the part, including only the data necessary for the desired level of quality and performance. During this process, the engineers should use the supplier as a consultant and provide drafts of the specs for review.

TQC and Supplier Support

The first step in bringing a supplier into the Just-In-Time system is to start a TQC program. This program should raise the quality level of the supplier until the company no longer needs to do receiving inspections of its products. The TQC effort must result in a near-perfect quality record on deliveries at the receiving dock.

In general, the higher the initial quality, the more difficult to raise the quality to a new level. Reaching near-perfection will require a lot of help from the manufacturing and design engineering groups. The goal is not only to help the supplier improve quality but also to shorten its lead time, for Just-In-Time requires a short response time from suppliers.

The manufacturing engineering people should concentrate on working with the supplier to improve its production process. The design engineering people should work with the supplier to improve the product's design to achieve high quality. They also should make sure there are no superfluous requirements in the product's specs that could hinder the supplier from reaching the new quality goals.

Obviously, the investment of considerable resources to improve the performance of suppliers is typically required. Such an investment provides an additional reason to develop long-term relationships with a small group of suppliers, for much effort will be wasted by supporting many suppliers or by switching suppliers frequently.

10.5 ENGINEERING AND JUST-IN-TIME EDUCATION

There is a general belief that Just-In-Time mostly involves manufacturing and is concerned very little with other departments. In practice, the effort to implement Just-In-Time depends very much on the support of engineers from many departments.

Technical support staff are needed to solve problems on the production line as soon as they occur. They are needed to work with suppliers to improve the quality of their products. They are needed to process engineering changes promptly. Finally, they are needed to produce a documentation set that clearly specifies the products that suppliers must deliver.

In order to provide the necessary support, the technical support staff need to understand the urgency of responding and to have a sense of priorities. The most effective way to achieve this is through a Just-In-Time training program.

It is recommended that engineers in all the relevant departments be trained at the same time as the manufacturing employees. It is also recommended that the training classes cover all the aspects of Just-In-Time, not just technical issues. This will help the engineers to understand the system as a whole and why their support is needed. Besides, the training will help to give them a sense of ownership regarding the program.

Thinking Simply and Designing Simply

Just-In-Time principles are applicable to product development as well as manufacturing. Just-In-Time means simplicity and no waste. Time waste is also a target. There is currently a tendency for designs to be overly complex and difficult to manufacture, which contradicts the Just-In-Time philosophy. The result could be called a *design waste*, since a designer who designs a product with more features than users need is wasting effort, time, and materials.

In the past, many fine products with excellent market acceptance have been changed in further design releases to include additional complex and difficult-to-use features, sometimes becoming less reliable as a result. Such changes are a blessing for the competition.

The definition of product features starts in the marketing department. But in most technology-driven companies, the product development staff have a strong say regarding features and the technologies to use in production. It is important for the staff to keep in mind the distinction between what is practical and what is technical waste. According to Just-In-Time principles, it is best to strive for simplicity in design. A simple product with the right features and of excellent quality will be eminently marketable.

10.6 SUMMARY

Normally when people start thinking about implementing a Just-In-Time system, the first thing that comes to mind is to send manufacturing managers to Just-In-Time seminars. This is correct as far as it goes. But it is also correct to send product development managers to the same seminars. When manufacturing managers later start talking about making changes in documentation and asking for support, the product development staff will understand the reasons for the requests.

Normally, design engineers tend to view such changes with skepticism. They generally respond slowly to changing documentation structures and part specs to make them simpler and less restrictive, for they have other priorities that interfere. Understanding the impact of Just-In-Time on the company will help to rearrange priorities and make engineers more willing to do certain tasks. Management support is critical for this change of attitude.

Just-In-Time requires active participation and interdepartment cooperation. Just-In-Time demands constant and unending improvement. Just-In-Time is everybody's business, including the product development staff. Without product development, Just-In-Time cannot be successfully implemented.

REFERENCES

FEIGENBAUM, ARMAND V. "New-Design Control." In *Total Quality Control*. 3d ed. New York: McGraw-Hill, 1983.

HALL, ROBERT W. "Product Design and Stockless Production." In *Zero Inventories*. Homewood, Ill.: Dow Jones-Irwin, 1983.

HAYES, ROBERT H., and STEVEN C. WHEELWRIGHT. "Link Manufacturing Process and Product Life Cycles." *Harvard Business Review*, January–February 1979, pp. 133–140.

———. "Matching Process Technology with Product/Market Requirements." In *Restoring Our Competitive Edge*. New York: John Wiley & Sons, 1984.

HAYNSWORTH, H. C., and R. TIM LYONS. "Remanufacturing by Design: The Missing Link." *Production and Inventory Management* (American Production and Inventory Control Society) 28, no. 2 (1987): 24–28.

SCHONBERGER, RICHARD J. "Design Leverage." In *World Class Manufacturing: The Lessons of Simplicity Applied.* New York: The Free Press, 1986.

WATTS, FRANK. "Engineering Changes: A Case Study." *Production and Inventory Management* (American Production and Inventory Control Society) 25, no. 4 (1984): 55–62.

11 ★

Just-In-Time Training Programs

Once a company starts to implement Just-In-Time, it needs to rally support from practically every level of manufacturing. It will also have to draw support from other departments. The most effective way to ensure success is to make certain that everyone involved learns the Just-In-Time principles. This requires instituting an intensive training program.

In practice, the Just-In-Time training program needs to be tailored to those attending the classes. The goal is for the trainees to understand the Just-In-Time principles and to know what the new system will demand from them. For example, upper management will want to hear what kind of improvement Just-In-Time is going to make in inventory turns, overhead reduction, and quality levels. Middle managers will express concern about line shortages and the reduction of the number of suppliers. They will also worry about the cooperativeness of suppliers regarding frequent deliveries and the additional workload required to support lot size reductions. The product development staff will be wary of the amount of support required and the possibility of constant interruptions when problems in the production line arise. They will also worry about the additional workload involved in bringing suppliers up to TQC quality standards. Finally, the workers in the production line will wonder what Just-In-Time is all about and whether it means more work for them or a reduction in employees.

The most effective way to address all those questions and concerns is to prepare a comprehensive training program. This program should reach all levels of management as well as the workers on the production line. The program should explain the Just-In-Time philosophy and the benefits the system can produce.

Everyone in the program should be made to understand that Just-In-Time cannot succeed unless they all work together. They should also understand that the system will not cure all problems overnight and that it will require hard work and dedication. Finally, they should understand that Just-In-Time is a continuous process and there is always room for improving the manufacturing process.

11.1 THE JUST-IN-TIME TEAM

The first step in implementing Just-In-Time is to assemble a team to act as the driving force behind its implementation. The team should be composed of individuals who are in a position to make a substantial contribution to the program. For example, it must include staff from the following departments: materials, purchasing, manufacturing support, product development, quality control, and manufacturing. The first priority of the team is to attend several Just-In-Time seminars to become familiar with the system. It is important that they are presented with case studies and are able to share the experiences of other companies. The team should also have access to literature on the subject.

The training of the team should continue until they become true believers. There cannot be a successful Just-In-Time system if those in charge of implementing it don't believe in the system. In fact, they must believe in it almost with a religious intensity.

Once the team members understand and accept the concepts, the next step is to prepare three sets of presentations. These presentations will introduce the Just-In-Time system to four levels in the company: upper and middle management, technical staff, and line workers. It is helpful to include video tapes showing case studies. A picture of a functioning Just-In-Time system is worth a thousand words.

The next step is to present an outline of the Just-In-Time plan. This plan must include clear goals for every aspect of the program. Make sure that all the team players agree with the goals and accept them as important. This will help to keep everyone motivated.

11.2 UPPER MANAGEMENT EDUCATION: THE FIRST PRIORITY

Creating senior management support for the Just-In-Time program is of the utmost importance. Start with a presentation that explains Just-In-Time fundamentals. During the presentation, outline the execution phases and the benefits to be expected from the system. Be careful not to represent Just-In-Time as the solution to all problems, for this might set expectations too high.

The presentation should include the basics of Just-In-Time, possible benefits, and a realistic time frame for realizing them. Also, include in the presentation a description of the resources and capital investment required to set up the system. This description will essentially be an outline of the plan to implement the system.

There is one area where senior management can directly contribute to the Just-In-Time program, namely, in dealing with reluctant suppliers. It is likely that some suppliers will be reluctant to accept the idea of short lead times and frequent deliveries of small lots. Senior management could contact a supplier's senior management to convince them to accept the new system.

Periodically reviewing the program's progress with upper management is recommended. The reviews should be realistic and present both successes and setbacks. Keep in mind that it will take time to get the system off the ground. Also, sometimes the process of implementation will seem to be moving backward instead of forward. Honesty about progress will add credibility to the effort and gain needed time and support.

11.3 MIDDLE MANAGEMENT EDUCATION AND INVOLVEMENT

Middle managers are the drivers of the Just-In-Time system. They must clearly understand the concepts and realize how much effort will be required to get the system implemented. Middle managers also must realize that Just-In-Time won't produce major improvements overnight and that it will take time, effort, and dedication to get results. They must be the most fervent supporters of Just-In-Time. Educating them about the system's fundamental principles should never end.

There is another area in which middle managers can provide a valuable contribution. In practice, it will be too costly to send every worker to Just-In-Time seminars outside the company. Middle managers, after they have been trained, could conduct training sessions within the company to introduce others to Just-In-Time principles. Not only will this help educate employees, but it will also demonstrate the commitment of the managers to the system. Periodic training sessions should also be conducted as the implementation progresses.

Middle managers must truly embrace Just-In-Time principles. No effort should be spared in convincing skeptical managers to believe in the Just-In-Time system. Just-In-Time will only succeed through the constant efforts of true believers to improve the system one step at a time.

11.4 WORKER TRAINING

When introduced to Just-In-Time, production workers will express concern about the impact of the new system on their jobs. They will ask whether it means a reduction in the work force or an increase in the workload. These issues must be addressed in the presentation, and the workers must be made to understand the principles of the system and the improvements it will bring to their jobs. Presentation topics should include the impact of the system on production linearity, small lots, quality improvements under TQC, and line-stopping authority. Also discussed should be the secondary benefits, such as reductions in space, materials stored on the production line, and paperwork.

In general, workers will not be concerned about increases in inventory turns and reductions in overhead. These are not among their responsibilities. As in the case of managers, it is important not to set the workers' expectations too high or lead them to believe that Just-In-Time will produce quick results. If it is made clear that success will require time and dedication, the workers will be more inclined to maintain their faith in the system.

One way to make the presentation interesting to the workers is to set up simple examples of the way Just-In-Time works. Another is to show video tapes of case studies. There are plenty of appropriate video tapes on the market.

It is imperative to conduct more than a single training session. Rather, there should be periodic reviews and requests for feedback on progress and for suggestions for improvements. To prove to workers that their input is of value, any reasonable suggested changes should be instituted. This will encourage them to continue providing feedback, which what Just-In-Time is all about—continuous small changes that add up to big improvements.

Finally, it is advisable to balance the presentations equally between Just-In-Time and total quality control. Workers must keep in mind both quality and simplification.

11.5 DEVELOPING A TRAINING PROGRAM

A good training program never ends but keeps evolving in parallel with the implementation of the Just-In-Time system. As it evolves, it should include accomplishments, such as the solutions of problems met along the way. The idea is to learn from mistakes so as not to repeat them.

Training should be in the hands of a set of true believers. They must be dedicated and experienced and have an established track record, which will provide them with credibility.

The training process should be simple, with lots of practical examples. Also, the classes should provide opportunities for feedback as the implementation progresses. It is advisable not to stop the training program after the initial phase. Refresher classes are very important even for the most experienced groups.

One approach is to invite groups within the company that are ahead in the implementation of Just-In-Time and have them present case studies and discuss major problems they have solved. Nothing will help a group newly implementing Just-In-Time more than listening to the experiences of other groups.

The importance of training cannot be overestimated. A Just-In-Time system changes the way that people perform their work and it represents a drastic departure from well-established principles of operation.

Materials department staff and manufacturing managers are conservative by nature. They want to have safeguards to ensure the shipment of products. Also, manufacturing failures are very visible, for there won't be any end-of-the-month shipments. Therefore, manufacturing managers are very slow to accept changes that would put their jobs at risk. The only way to overcome their conservatism is to make sure they are aware of the benefits of Just-In-Time and recognize that in the long run their jobs are going to be simpler. This is best done by putting in place a good training program.

11.6 SUPPLIER TRAINING

Not only company employees but also suppliers must be educated about the fundamentals of the Just-In-Time system. In general, suppliers will have the same concerns about the system as the materials department staff. Small lots, frequent deliveries, and changes in the schedule based on demand will cause them the same worries. To ease their concerns, the company should treat suppliers in the same way as company staff. Using training classes as a vehicle, the company should sell the advantages of the new system as well as explain its principles.

The first step is to inform a supplier about the company's intentions to implement a Just-In-Time supplier program. This can be done by visiting the supplier and presenting the Just-In-Time principles and the supplier program plans. It is very important that the supplier's senior managers attend this presentation, for they are the only ones who can make a commitment to such a program. The presentation should cover both the Just-In-Time system and the TQC program (see Chapter 6 for details).

One important factor to consider is that a supplier will have to do the same selling job with respect to its employees. The company can help the supplier by sharing its training program. This can be done very simply by giving training materials and copies of video tapes to the supplier and by training its trainers.

As with other training programs, it is beneficial to follow up the training with periodic reviews. Sharing is the key here; the more the company shares with the supplier, the more progress the company is going to make.

11.7 DEVELOPING SYSTEMS AND PROCEDURES

Once the company has a plan to implement Just-In-Time, it should develop the systems and procedures that determine the functioning of the system. The systems and procedures define the operational roles, responsibilities, and authority of every group involved. The Just-In-Time system will not succeed if it is not properly defined on paper, for employees will otherwise deviate from the main procedures and the result will be disorganization and even chaos.

There are two key points to remember when putting the systems and procedures in writing. One is to avoid making the procedures lengthy and difficult to understand. Employees will ignore lengthy procedures, and furthermore they are a waste of time and paper.

The second point to consider is that procedures must be reviewed with those who are going to use them. The system won't work unless the employees understand the procedures and consider them practical. It is important to get feedback on the procedures and modify them accordingly. Once this has been done, a series of training classes will help everyone understand his or her role in the new system.

The procedures should not be cast in concrete. After a period of time, perhaps three to six months, there should be a procedures review meeting with those who are using them. Corrections should be made on the basis of the feedback. This process of review and correction can be repeated indefinitely.

11.8 SUMMARY

The success of a Just-In-Time training program depends on tailoring the program to the different groups attending the classes. These groups have different participation levels based on their responsibilities in the company. Those in upper management are the supporters of the program. Their involvement provides leadership and inspiration to change things. Those in middle management are the drivers of the Just-In-Time system. They must learn to the last detail the benefits of the system. They are also responsible for making things happen and must clearly understand the process. The workers and support staff are responsible for running the Just-In-Time system. They are in the front lines and must become intimately familiar with the operational details of the system. This familiarity can be largely achieved through the use of numerous practical examples.

Another factor to consider is the need for feedback and for sharing experiences throughout the initial phase. After the first sessions, the training classes must include feedback from the people involved in implementing the system. Also, groups ahead of other departments should share their experiences and the results they have achieved. This will keep morale high.

Finally, a Just-In-Time training program should be never-ending. A Just-In-Time system demands perpetual vigilance, for new forms of waste will eventually appear. The solution is to continue training and training and always to be looking for new ways to improve further what has already been improved.

REFERENCES

ANDREW, CHARLES G. "Motivation in Manufacturing." *Production and Inventory Management* (American Production and Inventory Control Society) 27, no. 2 (1986): 133–142.

BELL, ROBERT R., and JOHN M. BURNHAM. "Managing Change in Manufacturing." *Production and Inventory Management* (American Production and Inventory Control Society) 28, no. 1 (1987): 106–115.

DAVIS, HAROLD S. "Management: What We Can Learn from the Japanese." *Production and Inventory Management* (American Production and Inventory Control Society) 27, no. 1 (1986): 85–89.

DYER, WILLIAM G. *Team Building: Issues and Alternatives*. Reading, Mass.: Addison-Wesley, 1977.

LEVINE, MARC. "Making Group Collaboration Work." *Production and Inventory Management* (American Production and Inventory Control Society) 28, no. 3 (1987): 31–33.

LOUNGE, DICK. "JIT Manufacturing: The People Aspect." In *EMTAS '86: Conference Papers*. Dearborn, Mi.: Society of Manufacturing Engineers, 1986.

SUZAKI, KIYOSHI. "JIT and Worker Participation." Paper presented at APICS Annual Conference, Toronto, 1985.

12 ★

A Blueprint for Just-In-Time

The previous chapters have focused on the Just-In-Time philosophy and how the Just-In-Time system impacts on the operational environment of a traditional manufacturing organization. The question still to be answered is how to implement such a system.

This chapter offers a guide to preparing a Just-In-Time implementation plan. Please bear in mind that what is being offered is only an outline and it does not cover all possible cases. Every company, product, and factory is unique in certain ways. A Just-In-Time program requires a certain degree of customization to effectively address individual needs. What follows is only a framework for constructing an actual plan.

12.1 PRACTICAL ADVICE AND MISTAKES TO AVOID

The first step in implementing a Just-In-Time system is to understand where the company is at the present time. This might seem simple, but in some cases it is very difficult. It is advisable to take inventory of the present system before making any changes in it. For example, how many suppliers are being used now? How many of them are supplying the same type of parts? What are their actual lead times? What are their rejection rates at the receiving inspection stations? What is their ranking by dollar contribution to purchases?

The next step is to understand the company's inventory. For example, how many days, weeks, or months of inventory are at hand for every product line? How much of that inventory is on the production line? How much of the WIP is not active? What is the linearity of the production process? Does the company build at the end of every month or every quarter?

How many work orders have been released to the manufacturing floor? How many are still open and past due? Is the production process on paper? How much manufacturing floor space is being used? How much shelf space is on the production line? What is the quality level when the product arrives at the customer's site? How much rework is done?

Answering these questions is equivalent to getting a medical checkup. The information will provide an idea of the current health of the organization. It will also provide a set of reference points to monitor progress during the implementation of the Just-In-Time system. The company should not expect instant progress on every front. Therefore, it must select the most problem-ridden areas to work on first, then go on from there.

Note that in some cases there is a certain order in which Just-In-Time implementation steps must be done. Be careful not to jump ahead and execute a step without having completed the prerequisite steps. For example, workers should not stop the line for problems until the proper support system is in place. The continual line stopping will demoralize them and erode their confidence in the new system. Again, do not stop doing receiving inspections until the suppliers are participating in TQC programs and their quality levels are acceptable. Stopping receiving inspections too soon will result in bad material reaching the line. The important thing is to understand the correct sequence of tasks. Then make sure that those in charge of implementing the system are following the process correctly.

12.2 STARTING SMALL

Let's assume that after reading this book you believe that Just-In-Time is the right way to go. You call a group of managers into a meeting to start talking about the new system. At that moment, you take a look at your company and realize that the task is enormous—it employs numerous people and produces numerous products. You also realize that the company has a commitment to ship at the end of every month that cannot be compromised. The best advice for someone in your situation is to *start small.*

During the early Just-In-Time implementation meetings, you can brainstorm to decide which departments or which products to apply Just-In-Time to first. The tendency is to select an easy product, one which could enter the Just-In-Time system without disrupting the current system. The reasoning is that the company needs some Just-In-Time experience before tackling a more complicated job.

On the contrary, you should select a product that is already in trouble. Not only will more be learned but the chances of making the situation worse will be reduced.

In summary, the best advice is to pick a line that is small and troublesome. Then, as you learn to solve problems in implementing Just-In-Time, you can apply this knowledge to bigger lines and more complex products. This approach will produce no waste of time—an appropriate achievement for a Just-In-Time system.

12.3 JUST-IN-TIME AND SMALL PRODUCTION RUNS

A low production rate is the most common excuse for avoiding the implementation of a Just-In-Time system. In fact, experience shows that the best time to start implementation is when a production line is nonexistent or when the manufacturing rate is small. Problems will be much smaller and easier to correct than when there is a high volume. If a company waits until it reaches a certain size, it will miss a great opportunity to do the job right from the beginning.

To expand on this point, the best way to implement Just-In-Time in a large organization is to apply it to small areas or processes one at a time. If necessary, divide large areas into smaller ones.

The same principles apply to small production runs that will never get bigger. Remember that the optimum lot size for a Just-In-Time system is one unit.

12.4 JUST-IN-TIME CHAMPIONS

Just-In-Time involves a war against waste, and, as in any war, there has to be an army to fight the battles. In this case, the workers and the support staff constitute the army and the team in charge of implementation is the generals. No army is complete, however, without a supreme commander—a champion.

The champion is the person who is going to pick up the Just-In-Time banner and never let it fall, the person who will move mountains, cross interdepartmental barriers, and lead the troops in the never-ending battle against waste.

A Just-In-Time program needs such a champion from its very beginning. This person must be high enough in the organization to carry substantial clout. Typically, the champion should be a director or a vice president. But being at the right level is not enough. The champion must not be a mere figurehead. The champion must be a driving force, an inspirational leader, someone who makes things happen.

The champion should take the same training classes as the core team and become intimately familiar with the Just-In-Time concepts. The champion should be part of the team and participate in the decision-making process. However, the champion should not make all the decisions during the process of defining and implementing the system. Just-In-Time requires spreading the responsibility for decision making down through the organization. The champion's role is to be the inspirer, the strategist, the leader.

12.5 FRAMEWORK FOR A JUST-IN-TIME SYSTEM

This section presents an outline of a Just-In-Time program. This outline is broad enough to fit most organizations, but keep in mind that every program will require some customization to meet particular needs. Think of the outline as a road map of Just-In-Time territory. The ultimate selection of the roads to take will of course depend on individual needs.

Phase 1: The Just-In-Time Team and the Education Program

In this phase, the core team is assembled and the Just-In-Time education process is started. It is also important to recruit a champion who will go through the training process. Training should be intensive and take from one to three months. The goal is to make the team into true believers. Only then should they draft the Just-In-Time implementation plan.

This plan should include a companywide training program in the principles of Just-In-Time. It should also include the product line for the program and the set of goals to be achieved. The employees who will be involved should have their responsibilities outlined in the plan. Finally, the plan should include a time frame for implementation and a list of critical suppliers that are intended to become Just-In-Time suppliers.

One critical decision is the selection of a repetitive manufacturing software package. It is also critical that the team determines the changes required to set up a Kanban system and to modify the cost accounting system so it can operate under Just-In-Time.

Phase 2: Initial Implementation on the Production Line

During this phase, the repetitive process to build the product should be defined. This requires changing the structure of the bill of materials to support a Just-In-Time system. Simultaneously, the manufacturing engineers should do a new layout of the factory with the goal of reducing floor and storage space.

The management information system department should start installing the software modules to handle the repetitive process. The Just-In-Time team should define the systems and procedures to run such a program and then review them with everybody involved. The key items to define in this step include the production line capacity and the desired daily production rates.

The repetitive process flow should define the way the stockroom will issue the daily material. For example, it should define which material is bulk and issue it for longer periods. The manufacturing engineering staff will define the containers to carry the daily material, the quantities (based on daily rates), and the proper routing to the work centers (the goal being to minimize the idle time of workers).

The manufacturing engineers should complete a study of process lead times and line balancing before they convert the process to repetitive manufacturing. This will uncover bottlenecks they must correct ahead of time.

Phase 3: Implementation of a Total Quality Control Program

The implementation of a TQC program is as critical as the previous steps. A TQC program should be started in parallel with the Just-In-Time program, for neither program will succeed without the other.

A TQC program is primarily the responsibility of the quality control department. But for a successful implementation, the department needs the support of the manufacturing and engineering staffs. In general, a TQC program affects the factory on the one hand and the suppliers on the other.

The implementation of internal TQC should take priority. Not that the suppliers should be ignored, but the most important goal is to increase quality within the factory, at least initially. The first step is to define the quality process of the production line, then to recruit and train the quality teams in charge of solving the problems on the line. During this phase, a simple data-collecting system should be developed to gather information on quality problems. The information will be used to prioritize the problems. The quality team can then be directed to deal with the problems in the order of their priority.

The TQC plan should include the training of the workers as inspectors. It should also include arrangements for removing quality inspectors from the line. Once they are removed, it should be very clear that the responsibility for the quality of products rests completely on the shoulders of the workers building them. This is the heart of the TQC program.

To speed the transfer of the responsibility for quality from the inspectors to the workers, there should be a good cross-training program for every worker on the line. Also, there should be a simple way to measure the quality of the product throughout the process and a very clear set of goals that everybody understands.

In tandem with the above, a system should be developed to collect data on potential Just-In-Time suppliers, including data on part problems discovered during receiving inspections and during the production process as well as data on the parts once they reach customers as components of final products.

Once the internal TQC process is in place and data have been collected on the suppliers, it is time to begin working with the suppliers to improve the quality of their parts. A TQC plan should produce a quality-certified group of suppliers that will compose the initial set of Just-In-Time suppliers. This group should now be able to start shipping parts frequently and in small lots. At the same time, the company can eliminate the receiving inspections and take the parts directly to the production line, thereby dispensing with the need for buffer inventories.

Phase 4: Conversion of the Production Line to Just-In-Time

This is the phase in which the production line is converted to a repetitive process. The materials planners will close the work orders in the computer and redistribute the materials to match the target daily rate. Also, the stockroom starts to issue materials in a repetitive way and the backflush points become effective. From now on the factory completions will pull the materials, and the stockroom will issue them only when there is a demand from the production line. The next step is to record completions in the factory daily and make every worker aware of the daily goals.

Those involved in the Just-In-Time system need some latitude until they start producing results. In fact, initial mistakes and low output should be expected. Contingency plans should be devised to ensure that shipping commitments are met.

To safeguard the quality of the product now that the line inspectors have been eliminated, it is recommended that a thorough final inspection be done before shipping. The Just-In-Time goal is to reduce final inspections as the quality of the product improves. At the beginning, however, thorough final inspections are necessary for peace of mind. Final inspection yield reports will indicate when this policy can be relaxed.

At the end of this phase, much has been achieved: The factory has a new layout; materials are issued only on demand and in a repetitive way; a Kanban system has been implemented for some line products; a new cost accounting procedure has been put in place to capture material and labor; and the line workers are their own inspectors. In short, the Just-In-Time system has been fully launched.

Phase 5: Working with Suppliers

The company has less control over the changes that occur during this phase than in the case of the previous phases. Up until now, the changes have involved departments that are under the management of people who want to implement Just-In-Time. Phase 5 is completely different. The company must work with suppliers who require a great deal of motivation, for they are being asked to alter radically the way they do things. The company will appear to be requesting them to increase their workload in order to satisfy its needs.

Phase 5 requires patience, negotiation, and sometimes tough decisions, for a non-responsive supplier might have to be replaced.

Chapter 6 contains guidelines for establishing Just-In-Time supplier programs. Preliminary tasks include creating a list of suppliers prioritized in order of their contributions to the products and collecting information about their quality levels, actual lead times, and delivery track records.

Success in this phase depends on making suppliers into believers in the Just-In-Time system. They will be required to improve the quality of their parts (eliminating the need for inspection) and to ship small lots frequently.

It is possible during this phase to develop an in-plant store program with one or several suppliers, thus creating a true partnership.

Phase 6: Assessing Just-In-Time Performance

As noted at the beginning of this chapter, the company must analyze itself and its relationships with suppliers before beginning to implement a Just-In-Time system. By the time it reaches phase 6, it has probably used the system for nine months to a year and spent about a year working with key suppliers to get them into the system. At this point, the company should again analyze itself and again ask the questions listed in Section 12.1. It should then compare the results of the analysis with the results before. The differences will indicate the level of success achieved.

Two outcomes are possible. First, the achievements might not be satisfactory enough to justify the effort. In this case, the problems need to be understood and corrected. Second, the goals outlined at the beginning of the program might have been accomplished, in which case the implementation has clearly been successful.

In either case, it is recommended the company set new goals and start working to achieve them. There is no end to the improvements that can be made.

12.6 THE NEVER-ENDING PROCESS OF IMPROVEMENT

It bears repeating that Just-In-Time is a never-ending process. So is the process of improving quality and developing good supplier relationships. In a manufacturing organization there is always room to improve, to learn from experience, to do things better the next time. All that is needed is creativity and persistence.

Just-In-Time, like any other discipline, has a learning curve, and the company can learn from experience how to do things better the next time around. That is the reason why it is so important to continue the program once the original goals have been achieved. The only difference as the program progresses is that the goals set will be more difficult.

12.7 SUMMARY

There are a few mistakes to avoid during the implementation of a Just-In-Time system. One is to devise an ambitious plan without allocating adequate resources to do the job right. In this case, the company will be risking failure by setting expectations too high without the proper support. Also, the credibility of the program will be undermined.

On the other hand, slow or no progress will also damage the credibility of the program. One way to avoid both problems is to balance resources and expectations and to track progress carefully. Deviations from the main plan of action should then be corrected expeditiously.

One way to monitor progress is to have periodic reviews at every level in the organization. Sometimes managers don't have enough direct contact with workers to evaluate progress. If that is the case, the company must set up a feedback system to detect and solve problems quickly. It should also institute a policy of accepting honest mistakes and providing help to avoid their repetition. This will encourage honest performance on the part of workers.

A Just-In-Time system is going to uncover many long-standing problems. The system should not be blamed for these problems. Indeed, one of the system's best features is that it makes problems visible as soon as they occur.

It is very important not to compromise Just-In-Time principles during implementation. The company should stick by the book. Each rule is vital to the survival of the system and should be enforced scrupulously.

REFERENCES

CHIANG, WILLIAM. "Principles of Planning for JIT Production Lines." In *EMTAS '86: Conference Papers*. Dearborn, MI.: Society of Manufacturing Engineers, 1986.

COOPER, CARL. "The Japanese Connection: Imitate or Emulate?" *Production and Inventory Management* (American Production and Inventory Control Society) 25, no. 3 (1984): 114–125.

GODDARD, WALTER E. "The Continually Improving Process." In *Just-In-Time: Surviving by Breaking Tradition.* Essex Junction, VT: Oliver Wight Limited Publications, 1986.

HALL, ROBERT. "Kawasaki U.S.A. Transferring Japanese Production Methods to the United States: A Case Study." American Production and Inventory Control Society, 1982.

———. "Implementing Stockless Production." In *Zero Inventories.* Homewood, Ill.: Dow Jones-Irwin, 1983.

HAYES, ROBERT H., and STEVEN C. WHEELWRIGHT. "Managing Changes in Manufacturing's Technology and Structure." In *Restoring Our Competitive Edge: Competing through Manufacturing.* New York: John Wiley & Sons, 1984.

PLENERT, GERHARD. "Are Japanese Production Methods Applicable in the United States?" *Production and Inventory Management* (American Production and Inventory Control Society) 26, no. 2 (1985): 121–129.

SCHONBERGER, RICHARD J. "Managing the Transformation." In *World Class Manufacturing: The Lessons of Simplicity Applied.* New York: The Free Press, 1986.

13 ★

The Challenge of Just-In-Time: Making the Complex Simple

Just-In-Time is a revolutionary concept that challenges by its very simplicity. It introduces no advanced technology or complicated principles but instead strives to eliminate the unnecessary burdens of complexity.

The advent of technology and the increased complexity of products have driven companies to complicate manufacturing processes. But complicated processes are not necessary, as Just-In-Time demonstrates. A company can build the same product with less labor, less overhead, more quality, and on time by using the principles of the Just-In-Time system.

Making the complex simple is the main goal of Just-In-Time, not only in manufacturing but in other departments of a company as well. The returns that can be gotten from simplifying will be surprising.

The job of implementing Just-In-Time is never done. A Just-In-Time system will be in constant evolution, since understanding gained from experience can always be used to try to simplify further.

13.1 THE CHALLENGE OF PERFECTION

Just-In-Time means doing the job right the first time and permanently solving problems as soon as they appear. It also means efficient use of resources and fixing deficiencies in the system.

The change from business as usual to the pursuit of perfection does not occur overnight. The stimulus should come from the top and percolate down through the organization. The most effective way to motivate people to seek perfection is by setting the example.

Just-In-Time forces individuals to think practically about analytical matters. The goal is to reverse the trend toward complexity and instead make things simple and efficient. It is always possible to question whether there is a need to do something or whether there is a simpler way to do it. And even if certain simplications are made as a result, the question whether a still better and simpler way exists should be raised.

13.2 THE TWO MAIN PRINCIPLES

The two main Just-In-Time principles are that pull systems of manufacturing are preferable to push systems and that waste of any kind should be eliminated.

In a Just-In-Time system, the demand for material should start in the shipping dock. Then, throughout manufacturing, the preceding process should always pull material from the subsequent process. Just-In-Time is a pull chain that starts at the end of the process and reaches all the way to the suppliers. Do not confuse this pulling with the natural flow of materials in the process, which is in the other direction.

The second main principle is to eliminate waste. Elimination of waste is so important that it is reviewed in detail in the following section.

13.3 THE WAR AGAINST WASTE

Just-In-Time defines waste as anything that doesn't add any value to the product. In a traditional manufacturing operation, waste is commonly associated with scrap and rework. Just-In-Time, in contrast, has expanded the definition to include areas that are normally considered to be overhead. Furthermore, Just-In-Time extends the concept of waste to suppliers and the way they conduct their businesses. Companies normally have an arm's-length relationship with their suppliers and they are concerned with price, delivery, and quality in that order. Just-In-Time requires that a company becomes involved with a supplier's production process to reduce costs and improve quality and delivery performance. This entails that the company and the supplier form a kind of partnership.

Waste exists in many forms, and some are accepted as normal. Just-In-Time sensitizes people to the fact that waste is not necessary and that eliminating waste reduces the cost of products. Some forms of waste are very easy to detect and correct, but the subtle, hidden forms are very difficult to uncover. Below is a list of the kinds of waste present in many departments and suppliers. Not all kinds of waste are mentioned, but most of the important ones are.

WASTE IN THE PRODUCTION LINE

- Rework
- Poor workmanship
- Low yields

- Buffer inventories
- Line down due to equipment failure
- Absenteeism
- Long breaks
- Line down due to material shortages
- Engineering changes
- Additional assembly labor due to poorly designed product
- Lack of proper tools
- Unclear assembly instructions
- Poor training
- Poor factory layout
- Long machine setup times
- Low-quality raw materials
- Excess paperwork
- Scrap
- Worker idle time

WASTE IN THE MATERIALS DEPARTMENT

- Buffer inventories
- Excess materials
- Obsolete materials
- Incoming inspection of materials
- Excess freight or duty
- Inventory loss
- Too many suppliers
- Too many purchase orders
- Early or late shipments
- Large facilities for storing inventory
- Travel
- Phone expediting
- Receipt count discrepancies against purchase orders
- Wrong dunnage, boxes, or pallets
- No carrier plan
- Expediting shortages
- Poor forecast and planning of materials
- Too much bidding with suppliers
- Switching suppliers

- Paperwork
- Poor forecasts from sales department

WASTE INVOLVING SUPPLIERS

- Poor quality parts
- Early or late shipments
- Large shipments
- Shipment count discrepancies
- Rework
- Poor process yield
- High cost of sales
- Expediting
- Invoicing discrepancies
- Wrong freight
- Poor specification of product
- Overspecification of product
- Excess sales commissions
- Wrong dunnage, boxes, or pallets
- Poor forecast and changes in the production schedule

WASTE IN DESIGN ENGINEERING

- Poor documentation
- Marginal design
- Too many parts in design
- Too many different suppliers for parts used in design
- Complex design to assemble
- Complex design to test
- Part tolerances too loose or tight
- Complex manufacturing process required
- Poor testing before releasing to manufacturing
- Late release
- Parts and suppliers not used in previous designs
- Use of unreliable components
- High material cost
- Too many configurations in product
- Too many engineering changes and rework
- Too many bills of materials and levels

- Not designed for foolproof assembly
- Low reliability design
- Use of poor quality parts
- Design includes features customers don't need

From this list it is clear that waste can occur in many hidden forms. Just-In-Time calls for a war against waste, and everybody must share the responsibility of fighting whatever forms of waste are present. Once employees learn how to ferret out waste in its many forms, the results will be surprising. The Just-In-Time system will be in high gear when that happens.

13.4 TEN RULES TO REMEMBER

This section lists the most important rules for applying Just-In-Time. Following these rules will increase the chances of getting a Just-In-Time system off the ground. (Note that the list is not exhaustive and there are other rules that will also be helpful.)

1. Start working on a Just-In-Time system as soon as possible. Don't wait until the company has a large production volume. When that time comes, changes in the system will be more difficult. Pressure to ship products will always work against analysis and change.

2. Don't use the excuse of low volume to avoid implementing Just-In-Time. The system will work in any environment. Remember, the ideal volume in Just-In-Time production is one unit.

3. Use repetitive manufacturing, daily schedules, Kanban, and daily pull methods. Using these avoids the complexity of work orders and allows problems to be detected as soon as they occur.

4. Get senior managers involved immediately. Their support is important for the success of the program. Also, when you need to invest money in capital equipment, their help will make the process easier.

5. Start educating middle managers and workers about Just-In-Time principles immediately. Their understanding of Just-In-Time must be clear, even if they are initially skeptical of results.

6. Have senior managers get involved with key suppliers at the beginning of the program. This is essential for motivating the suppliers to support Just-In-Time. Point out to suppliers that they stand to gain from having a long-term relationship with, and being single-source suppliers of, the company. Remind suppliers that sometimes an initial investment in support is required.

7. Don't start with a global program. Choose a few key areas in which to implement Just-In-Time. Then pick another area. If you try to do too much, the system won't work smoothly and employees will get frustrated and lose interest.

8. Develop Just-In-Time systems and procedures at the beginning then provide training before instituting them. Don't leave the procedures for later. Otherwise the system will degenerate. As the organization gains experience, review the procedures and look for possible improvements.

9. Develop a set of measurable goals for the Just-In-Time program. Then monitor them and review their status with managers and workers. Allow the goals to be adjusted midstream if necessary.

10. When the goals are being consistently achieved, set new higher goals.

13.5 SUMMARY

Just-In-Time is a discipline that involves not only manufacturing but also many other areas of a company. The system pulls together all the different departments and closes up all the empty gaps.

Appendix B contains a performance table that can be used as a general reference in monitoring progress during the implementation of a Just-In-Time system. It is advisable to record the actual values of every item in the table before implementation begins. During implementation, compare the progress achieved with the goals suggested in the table.

Just-In-Time is a challenge to increase productivity in new ways. Specifically, it is a challenge to do more with less. That challenge now lies in front of you. Good luck!

REFERENCES

HANNAH, KIMBALL H. "Just-In-Time: Meeting the Competitive Challenge." *Production and Inventory Management* (American Production and Inventory Control Society) 28, no. 3 (1987): 1–3.

MAREK, E. WILLIAM. "Changes Worth Making: Managing the Mundane." *Production and Inventory Management* (American Production and Inventory Control Society) 28, no. 2 (1987): 104–109.

SUZAKI, KIYOSHI. "Corporate Culture for JIT." Paper presented at APICS Zero Inventory Philosophy and Practice Seminar, St. Louis, 1984.

WALLEIGH, RICHARD C. "What's Your Excuse for Not Using JIT?" *Harvard Business Review*, March–April 1986, pp. 38ff.

Appendix A

Metaphor JIT Supplier Agreement

Purchase Agreement Number ________

This Agreement dated as of ________ is entered into by and between ________________ (hereinafter called Supplier) and Metaphor Computer Systems, Inc. (hereinafter called Metaphor).

Whereas, Supplier is interested in manufacturing ____________________ to Metaphor drawings and specifications and selling such equipment to Metaphor; and whereas, Metaphor is interested in purchasing such equipment from Supplier:

Therefore, it is mutually agreed:

1. SCOPE

The scope of this agreement is to define a long-term partnership between Metaphor and the Just-In-Time Supplier, where the Supplier obtains a fair return on his investment, and Metaphor obtains the very best quality, delivery, and price. Throughout the execution of this agreement, Metaphor and the Supplier will strive to have a harmonious relationship that will develop into a partnership for the good of both companies.

2. PRODUCT PURCHASED

Appendix A* will include the specifications of the product(s) that Metaphor will buy from the Supplier. This specification must be clear enough to uniquely identify the product(s) and it shall be changed

*Supplier agreement appendixes would normally be attached to the agreement. They have not been reproduced here, since they depend entirely on the specific terms that will have been worked out between Metaphor and the particular supplier.

only upon written mutual agreement by both parties. The Supplier warrants that he possesses sufficient technical manufacturing capabilities to manufacture and test said products in accordance with the herein referred specifications.

3. LENGTH OF THE CONTRACT

The length of this agreement shall be specified in Appendix B. Appendix B will also specify the quantity of product(s) that Metaphor will purchase from the Supplier with an approximate monthly rate for the first six months of the agreement.

4. DELIVERY SCHEDULES

Supplier commits to deliver the herein specified product(s) frequently, in small lots, and upon Metaphor releases. The frequency of deliveries and quantities included are specified in Appendix C. Where necessary, the Supplier will commit to a buffer inventory to achieve frequent deliveries. This buffer inventory, whole parts or critical long lead components, will be mutually agreed upon with Metaphor. Metaphor, in turn, will make a firm commitment to accept those parts in future deliveries on a noncancelable basis. Appendix C will define the size of the buffer inventory or the inventory of critical long lead items required to execute frequent deliveries.

5. PRODUCT ROLLING FORECAST

Metaphor will give to the Supplier a monthly forecast that will cover a forward window of six months. This forecast will roll automatically to the next month every month. The changes to this forecast must be in general accord with the product volume specified in the agreement.

6. SHIPMENT RELEASES NOTIFICATION

Metaphor will reserve the choice of accepting delivery of the product based on the release schedule rates specified in Appendix C. Metaphor will also reserve the right to execute monthly releases each month, always meeting the rescheduling policies specified in Section 7. The monthly releases can be executed verbally, by a phone call, or through some hard record mutually agreed upon by Metaphor and the Supplier. In any event, Metaphor shall notify the Supplier of the next monthly release no later than the 15th day of the previous month.

7. RESCHEDULING POLICY

Metaphor will furnish the Supplier with two types of releases. One is the Product Rolling Forecast specified in Section 5, and the other is a firm release that shall cover a forward window of three months. The deadline to give the Supplier the firm release will be the 15th day of the previous month of the three-month window.

Schedules changes will be allowed according to the following table:

First month = +/– 10% of number of units forecasted per Section 5.
Second month = +/– 25% of number of units forecasted per Section 5.
Third month = +/– 50% of number of units forecasted per Section 5.

Any variations (increases/decreases) different from the above table must be mutually agreed between Metaphor and the Supplier.

8. SUPPLIER'S LEAD TIME AGREEMENT

Metaphor and the Supplier agree to work jointly to decrease the Supplier's lead time in producing the product specified in Appendix A. In cases where it is necessary for the Supplier to buy and stock an additional number of parts above the normal production ratios he is using, Metaphor will commit to accepting those parts on a noncancelable basis. This arrangement must be recorded in writing. Supplier must do his best to reduce the lead times to be able to conform to the agreement rescheduling policies in Section 7.

9. PURCHASE ORDERS RELEASE

Metaphor and the Supplier are to use the best method to record the agreement transaction in a purchase order. The purchase order could cover a blanket release with a total number and a shipment rate method or could cover individual transactions. Purchase orders will identify each part ordered by its part number, quantity required and price, and will set forth the requested delivery rate for each such part.

Upon receipt of any purchase order submitted in accordance with the paragraph above, Supplier will give Metaphor written notice of acceptance of Metaphor's purchase order within ten (10) days by signing and returning the acknowledgment of the purchase order.

10. PRICES AND BILLING METHODS

Metaphor agrees to pay to Supplier the price for parts as set forth in Appendix D to this agreement. Prices quoted in Appendix D will be firm for the term of this agreement.

Metaphor and Supplier shall agree to the method of invoicing to avoid a large influx of invoices following the frequent deliveries. A monthly invoice statement or other forms of group invoicing shall be considered and agreed between Metaphor and the Supplier. The agreed-upon invoicing method will be specified in Appendix E.

11. PACKING AND SHIPPING METHODS

Metaphor and Supplier will agree upon the packing and shipping methods of the parts procured under this agreement. The packing container will conform to Metaphor's Just-In-Time philosophies and will transport the product while meeting the quality standard called upon within this agreement.

In the event that the packing method requires any special tooling or expense, Metaphor will be agreeable to sharing the initial cost upon discussion and proposal by the Supplier.

12. PRICE REVIEWS AND COST REDUCTION PROGRAMS

Metaphor and the Supplier, having entered into this long-term business relationship, agree that the Supplier should have a fair return on his investment and Metaphor the very best price. It is incumbent upon the Supplier, as he improves his process and quality, to share the cost savings with Metaphor in the form of reduced prices.

13. PRODUCT WARRANTY

Supplier will guarantee the product for a period of ______________ starting with the shipping date of the parts. Supplier agrees to repair all parts free of charge during the guaranty period.

14. FAULTY MATERIAL RETURNS AND CREDIT PROCEDURES

Metaphor and Supplier will agree on a faulty material return and credit policy upon the signing of this agreement and will establish who is responsible for freight in each direction. The policy method will be specified in Appendix F.

15. QUALITY STATEMENT

Supplier shall deliver all parts to Metaphor conforming to the specifications attached in Appendix A. All parts shall be 100% defect free.

16. TOTAL QUALITY CONTROL PROGRAM WITH SUPPLIER

Supplier shall agree to enter into a Total Quality Control (TQC) program with Metaphor. The purpose of the TQC program is to resolve all quality issues with the parts delivered under this agreement so that Metaphor can eliminate all Receiving Inspection functions at Metaphor's facilities.

The definition and scope of the TQC program will be defined between the Quality managers of Metaphor and the Supplier.

Source inspection of the product will only be implemented for a transitional phase. Once the Supplier is JIT/TQC certified, there will be no need for source inspection.

Metaphor reserves the right to inspect Supplier's product and process occasionally on the Supplier's premises.

Metaphor will furnish to the Supplier monthly quality reports on the quality performance of the product. This quality report will include product yields on the Metaphor production line and a list of problems causing the yield to be less than 100%.

17. JUST-IN-TIME/TQC SUPPLIER CERTIFICATION

Metaphor will certify the Supplier to be a JIT SUPPLIER to Metaphor after the Supplier has signed this agreement, has entered into a TQC program with Metaphor, and is showing satisfactory quality levels. The certification will be issued in the form of a plaque that the Supplier can exhibit on the Supplier's premises. Metaphor will have a duplicate display on Metaphor's premises showing the Supplier's qualifications.

Metaphor will reserve the right to disqualify the Supplier if the quality of the product does not meet the quality levels specified in this agreement, or if the Supplier does not show a corrective plan to solve the quality issues under the TQC program.

18. METAPHOR/SUPPLIER TECHNICAL COOPERATION

Metaphor and the Supplier must commit their manufacturing technical people to working with the Supplier to achieve the quality levels specified in this agreement, and to improving the manufacturing process in order to reduce the product lead times.

The level of technical involvement must be mutually agreed upon by Metaphor and the Supplier. It must not interfere with either party's normal business operation and will guarantee the confidentiality of the information within both companies.

19. STANDARD TERMS AND CONDITIONS OF PURCHASE

Metaphor's standard terms and conditions of purchase are attached and are a part of this agreement.

20. NOTICE

Any formal notice given during this agreement must be in writing and sent by certified mail, return receipt requested, to the following representatives of Metaphor and the Supplier:

Metaphor: ______________________________

____

____

Supplier: ______________________________

____

____

21. CONTRACT EXTENSIONS

Ninety days before the elapsing date of this agreement, Metaphor and the Supplier will enter into negotiation for the extension of this agreement. In the event that the agreement is not renewed in that time, this agreement will elapse and cease after proper arrangements are made to consume special parts purchased in advance by the Supplier with Metaphor's previous written consent.

22. CONTRACT VIOLATIONS

In the event that an agreement violation is detected by either party, a notification in writing will be sent to the other party. The other party must answer and disclaim the violation in writing within thirty (30) days upon receipt of such notification.

23. SINGLE SOURCE COMMITMENT

Upon signing this agreement, and in keeping with the spirit of the Just-In-Time System, Metaphor will commit to use the Supplier as the single source supplier of the part(s) specified in this agreement.

BUYER: Metaphor Computer Systems
1965 Charleston Road
Mountain View, CA 94043

Signed: ______________________________

Title: ______________________________

SUPPLIER: ______________________________

____

____

Signed: ______________________________

Title: ______________________________

Appendix B

Just-In-Time Goal Performance Table

Just-In-Time Goal	Performance Achievement		
	Low	Medium	High
Inventory turns	up to 10	10 to 25	over 25
WIP in the process	2 weeks	1 week	1 day
Cycle time reduction	25%	25–75%	over 75%
Scrap and rework reduction	30%	30–80%	over 80%
Supplier base reduction	25%	25–50%	over 50%
Number of suppliers under Just-In-Time	25%	25–75%	over 75%
Quality improvement[a]	50%	90%	100%
Number of parts with no receiving inspection Dock-to-WIP	25%	25–75%	over 75%
Factory space reduction	25%	25–50%	over 50%
Factory output linearity	85%	85–97%	over 97%
Productivity increase[b]	25%	25–50%	over 50%
Overhead reduction	20%	20–50%	over 50%

[a]Quality improvement is measured by the percentage of reduction in the number of defects.

[b]Productivity increases can be measured in different ways, for example, factory output against the number of hourly workers.

Index